Erwin Dee Kord (Ed.)

Technika SH-Z625

Erwin Dee Kord (Ed.)

Technika SH-Z625

Tesco, Pixel, Digital Camera, USB mass-storage device class

Solv

Imprint

Publisher:
Solv is a trademark of
International Book Market Service Ltd., 17 Rue Meldrum, Beau Bassin, 1713-01 Mauritius
Email: info@bookmarketservice.com
Website: www.bookmarketservice.com

Published in 2011

Printed in: U.S.A., U.K., Germany. This book was not produced in Mauritius.

ISBN: 978-613-7-81905-0

Contents

Articles

Technika SH-Z625 1

Tesco 2

Pixel 28

Digital camera 33

AA battery 47

Zoom lens 50

Focal length 55

USB mass-storage device class 58

USB 63

Secure Digital 89

PictBridge 111

Audio Video Interleave 112

References

Article Sources and Contributors 115

Image Sources, Licenses and Contributors 119

Technika SH-Z625

The **Technika SH-Z625** is a 6 Mega pixel compact digital camera sold by Tesco. The camera is powered by two AA batteries and has a 3X optical zoom, and a 5X digital zoom. The lens has a focal length ranging from 5.8 to 17.4 mm. It has no optical viewfinder, but has a 2.4 inch TFT LCD screen. It has automatic focus and flash. A single (micro USB) connection is used both as a USB 2.0 interface and video out. Under USB the camera acts as a USB mass storage device giving access to the SD memory card. The USB connection also allows printing via PictBridge. The video-out connection is combined in the USB cable and is used for play-back of video clips (without sound) and slideshows of still pictures. Video clips are recorded with sound, using an AVI file format.

Tesco

Type	Public limited company
Traded as	LSE: TSCO [1]
Industry	Retailing
Founded	1919 (East London)
Founder(s)	Jack Cohen
Headquarters	Cheshunt, Broxbourne, Hertfordshire, England, United Kingdom
Number of locations	5,380 stores (As at Feb 2010) (*see table below*)
Key people	• David Reid (Chairman) • Philip Clarke (Chief Executive)
Products	Groceries, consumer goods, financial services, telecoms
Revenue	▲ £60.93 billion (2010/11)[2]
Operating income	▲ £3.81 billion (2010/11)[2]
Net income	▲ £2.67 billion (2010/11)[2]
Employees	472,000 (2010)[3]
Subsidiaries	Tesco Stores Ltd.
Website	www.tesco.com [4]

Tesco plc (LSE: TSCO [1]) is a global grocery and general merchandise retailer headquartered in Cheshunt, United Kingdom.[5] It is the third-largest retailer in the world measured by revenues (after Wal-Mart and Carrefour) and the second-largest measured by profits (after Wal-Mart).[6] [7] It has stores in 14 countries across Asia, Europe and North America and is the grocery market leader in the UK (where it has a market share of around 30%), Malaysia, the Republic of Ireland and Thailand.[8] [9] [10]

Tesco House, the Tesco head office, in Cheshunt, Broxbourne, Hertfordshire

The company was founded in 1919 by Sir Jack Cohen as a group of market stalls. The Tesco name first appeared in 1924, after Cohen purchased a shipment of tea from *T. E. Stockwell* and combined those initials with the first two letters of his surname,[11] and the first Tesco store opened in 1929 in Burnt Oak, Middlesex. The business expanded rapidly, and by 1939 there were over 100 Tesco stores across the country.[12] Originally a UK-focused grocery retailer, since the early 1990s Tesco has increasingly diversified geographically and into areas such as the retailing of books, clothing, electronics, furniture, petrol and software; financial services; telecoms and internet services; DVD rental; and music downloads.[13]

Tesco is listed on the London Stock Exchange and is a constituent of the FTSE 100 Index.

History

Tesco store at Kingston Park, Newcastle upon Tyne

Tesco store at Kulim, Malaysia

20th century

A Branch of Tesco built inside the Hoover Building in Perivale, London (now a listed building).

Jack Cohen founded Tesco in 1919 when he began to sell surplus groceries from a stall at Well Street Market, Hackney, in the East End of London.[14] The Tesco brand first appeared in 1924. The name came about after Jack Cohen bought a shipment of tea from *T. E. Stockwell.* He made new labels using the first three letters of the supplier's name (TES), and the first two letters of his surname (CO), forming the word TESCO.[11] The first Tesco store was opened in 1929 in Burnt Oak, Edgware, Middlesex. Tesco was floated on the London Stock Exchange in 1947 as Tesco Stores (Holdings) Limited.[14] The first self-service store opened in St Albans in 1956 (which remained operational until 2010, with a period as a Tesco Metro),[15] and the first supermarket in Maldon in 1956.[14]

During the 1950s and the 1960s Tesco grew organically, and also through acquisitions, until it owned more than 800 stores. The company purchased 70 *Williamson's* stores (1957), 200 *Harrow Stores* outlets (1959), 212 *Irwins* stores (1960, beating Express Dairies Premier Supermarkets to the deal), 97 *Charles Phillips* stores (1964) and the Victor Value chain (1968) (sold to Bejam in 1986).[16]

Originally specialising in food and drink, it has diversified into areas such as clothing, electronics, financial services, telecoms, home, health, car, dental and pet insurance, retailing and renting DVDs,[13] CDs, music downloads, Internet services and software.

Jack Cohen's business motto was "pile it high and sell it cheap",[17] to which he added an internal motto of "YCDBSOYA" (You Can't Do Business Sitting On Your Arse) which he used to motivate his sales force.[17] [18]

In May 1987 Tesco completed its hostile takeover of the Hillards chain of 40 supermarkets in the North of England for £220 million.[19]

In 1994, the company took over the supermarket chain William Low, successfully fighting off Sainsbury's for control of the Dundee-based firm, which operated 57 stores. This paved the way for Tesco to expand its presence in Scotland, which was weaker than in England. In 2006 Inverness was branded as "Tescotown",[20] [21] because well over 50p in every £1 spent on food is believed to be spent in its three Tesco stores.[22]

Tesco introduced a loyalty card, branded 'Clubcard', in 1995 and later an Internet shopping service. As of November 2006 Tesco was the only food retailer to make online shopping profitable.[23] In 1996 the typeface of the logo was changed to the current version with stripe reflections underneath, whilst the corporate font used for store signage was changed from the familiar "typewriter" font that had been used since the 1970s. Terry Leahy assumed the role of Chief Executive on 21 February 1997, the appointment having been announced on 21 November 1995.[24] [25]

On 21 March 1997 Tesco announced the purchase of the retail arm of Associated British Foods, which consisted of the Quinnsworth, Stewarts and Crazy Prices chains in the Republic of Ireland and Northern Ireland, plus associated businesses, for £640 million.[26] The deal was approved by the European Commission on 6 May 1997.[27] This acquisition gave it both a major presence in (and marked a return to) the Republic of Ireland and a larger presence in Northern Ireland than Sainsbury's, which had begun its move into Northern Ireland in 1995.

In 1997 Tesco and Esso (part of Exxonmobil) forged a business alliance that included several petrol filling stations on lease from Esso, with Tesco operating the attached stores under the Express format. In turn Esso operates the forecourts and sells their fuel via the Tesco store. 200 Tesco/Esso stores now exist across the UK.

21st century

The company was the subject of a letter bomb campaign lasting five months from August 2000 to February 2001 as a bomber calling himself "Sally" sent letter bombs to Tesco customers and demanded Clubcards modified to withdraw money from cash machines.[28]

In July 2001 Tesco became involved in internet grocery retailing in the USA when it obtained a 35% stake in GroceryWorks.[29] In 2002 Tesco purchased 13 HIT hypermarkets in Poland. It also made a major move into the UK convenience store market with its purchase of T & S Stores, owner of 870 convenience stores in the One Stop, Dillons and Day & Nite chains in the UK.[30]

In October 2003 the company launched a UK telecoms division, comprising mobile and home phone services, to complement its existing Internet service provider business. In June 2003 Tesco purchased the C Two-Network in Japan.[31] It also acquired a majority stake in Turkish supermarket chain Kipa.[32] In January 2004 Tesco acquired Adminstore, owner of 45 Cullens, Europa, and Harts convenience stores, in and around London.[33] In August 2004, it also launched a broadband service. In Thailand Tesco Lotus was a joint venture of the Charoen Pokphand Group and Tesco, but facing criticism over the growth of hypermarkets CP Group sold its Tesco Lotus shares in 2003. In late 2005 Tesco acquired the 21 remaining Safeway/BP stores after Morrisons dissolved the Safeway/BP partnership.[34] In mid 2006 Tesco purchased an 80% stake in Casino's Leader Price supermarkets in Poland. They will be rebranded into small Tesco stores.[35] In 2003, Tesco took part in a joint venture with O2 to form the Tesco Mobile mobile virtual network operator.

In 2007 Tesco was placed under investigation by the UK Office of Fair Trading (OFT) for acting as part of a cartel of five supermarkets (Safeway, Tesco, Asda, Morrisons and Sainsburys) and a number of dairy companies to fix the price of milk, butter and cheese. In December 2007 Asda, Sainsburys and the former Safeway admitted that they acted covertly against the interests of consumers while publicly claiming that they were supporting 5,000 farmers recovering from the foot-and-mouth crisis. They were fined a total of £116 million. Tesco, which maintains that it was not a part of the cartel, is still under investigation by the OFT.

Although profits were £1.9 billion for the first half of 2011, sales figures were the worst in 20 years due to shoppers switching to budget rivals.[36]

Corporate strategy

Potters Bar Tesco

According to Citigroup retail analyst David McCarthy, "[Tesco has] pulled off a trick that I'm not aware of any other retailer achieving. That is to appeal to all segments of the market".[37] One plank of this strategy has been Tesco's use of its own-brand products,[38] including the upmarket "Finest", mid-range Tesco brand and low-price "Value" encompassing several product categories such as food, beverage, home, clothing, Tesco Mobile and financial services.

Beginning in 1997 when Terry Leahy took over as CEO, Tesco began marketing itself using the phrase "The Tesco Way" to describe the company's core purposes, values, principles, and goals[39] This phrase became the standard marketing speak for Tesco as it expanded domestically and internationally under Leahy's leadership, implying a shift by the company to focus on people, both customers and employees.[40]

A core part of the Tesco expansion strategy[41] has been its innovative use of technology.[42] It was one of the first to build self-service tills and use cameras to reduce queues.[43]

To protect its brand image, and given its expansion plans in Thailand, Tesco has recently been employing a policy of launching defamation proceedings. In November 2007, Tesco sued a Thai academic and a former minister for civil libel and criminal defamation. Tesco is insisting that the two pay £1.6 million and £16.4 million plus two years' imprisonment respectively. They have been alleged to have misstated that Tesco's Thai market amounts to 37% of its global revenues, amongst criticism of Tesco's propensity to put small retailers out of business.[44]

Tesco's main advertising slogan is "Every little helps". Its advertisements in print and on television mainly consist of product shots (or an appropriate image, such as a car when advertising petrol) against a white background, with a price or appropriate text (e.g., "Tesco Value") superimposed on a red circle. On television, voiceovers are provided by recognisable actors and presenters, such as Barbara Windsor, James Nesbitt, Jane Horrocks, Terry Wogan, Dawn French, Ray Winstone, Neil Morrissey, Martin Clunes, David Jason, David Tennant and Kathy Burke amongst others.

Corporate social responsibility

Tesco has made a commitment to corporate social responsibility in the form of contributions of 1.87% in 2006 of its pre-tax profits to charities/local community organisations.[45] This compares favourably with Marks & Spencer's 1.51% but not well with Sainsbury's 7.02%.[46] Will Hutton, in his role as chief executive of The Work Foundation recently praised Tesco for leading the debate on corporate responsibility.[47] However *Intelligent Giving* has criticised the company for directing all "staff giving" support to the company's Charity of the Year.[48]

In 1992, Tesco started a "computers for schools scheme", offering computers in return for schools and hospitals getting vouchers from people who shopped at Tesco. Until 2004, £92 million of equipment went to these organisations. The scheme has been also implemented in Poland.[49]

Tesco in Evesham

Starting during the 2005–06 football season, the company now sponsors the Tesco Cup, a football competition for young players throughout the UK.

In 2009 Tesco used the phrase, "Change for Good" as advertising, which is trade marked by Unicef for charity usage but not for commercial or retail use, which prompted the agency to say, "It is the first time in Unicef's history that a commercial entity has purposely set out to capitalise on one of our campaigns and subsequently damage an income stream which several of our programmes for children are dependent on." They went on to call on the public "...who have children's welfare at heart, to consider carefully who they support when making consumer choices."[50] [51] Tesco's own labels for personal care and household products are moreover cruelty free - this means, they are not tested on animals - http://www.gocrueltyfree.org

In June 2011, Tesco announced that it was working with 2degrees Network to create an online hub as part of its target to reduce its supply chain carbon footprint by 30% by 2020.[52]

In September 2011 a Greenpeace report revealed that Tesco supermarkets in China were selling vegetables that contained illegal pesticides or at levels exceeding the legal limit. A green vegetable sample from Tesco turned up methamidophos and monocrotophos, the use of which has been prohibited in China since the beginning of year 2007.[53]

Advertising

Tesco have used many Television adverts over the years. In July 2007 a DVD containing adverts from 1977–2007 was given to all members of staff. Early advertising stressed cheap prices and how to keep "The cost of living in check." In 1977 an advert was made where a till showed the prices to many items such as "baked beans $12^1/_2$p".

Prunella Scales, as Dotty Turnbull arguing about Tesco prices

Tesco's most famous advert was "Checkout 82," which was made in 1982, where a till would have a receipt coming out of it with the prices on. This advert had synth pop music as the backing and people singing "Check it out, check it out".

Adverts in the late 90s had Prunella Scales as Dotty Turnbull, arguing about Tesco prices. In 2003, adverts showed items and shopping trolleys talking about Tesco. Late 2000s adverts have included many celebrities and celebrity voice-overs such as The Spice Girls and the voice of actors James Nesbitt and Jane Horrocks.[54]

UK operations

Stores

Tesco's UK stores are divided into six formats, differentiated by size and the range of products sold. These are shown below:

Tesco Extra

Tesco Extra stores are larger, mainly out-of-town hypermarkets that stock nearly all of Tesco's product ranges. Exceptions include Trowbridge, Stafford, Hexham, Kingston upon Hull, Stevenage, Chesterfield, Clay Cross, Alfreton, Leigh, Wigan, Grimsby, Bulwell, Galashiels, Slough, Bradford, Eastbourne, Yeovil, Cradley Heath, Burnley, and Leyland, which are in the heart of town centres. Cardiff (Western Avenue) and Cardiff (Pengam Green), Warrington and Walsall College are located in inner-city locations. The first Extra opened in 1997 in Pitsea. The 100th store opened on 29 November 2004 in Stafford. The number of these is now being increased by about 20 a year, mainly by conversions from the second category.

Tesco Extra, Commercial Street, Batley, West Yorkshire.

The largest store in England by floor space is Tesco Extra in Walkden, with 185500 square feet (17230 m^2) of floorspace.[55] In 2011, a Tesco Extra will be constructed in St.Helens adjacent to the new St.Helens RLFC Stadium. The largest in Scotland is the Silverburn store. The largest in Wales is at Parc Fforestfach, Swansea, which is 112000 square feet (10400 m^2) constructed in 2003. The 200th Extra store was opened in October 2010 in Bishop Auckland.

Other large stores include Bar Hill, Cleethorpes, Newcastle upon Tyne, Milton Keynes, Stockton-on-Tees and Watford, which are all in the 120000 square feet (11000 m^2) range. Newer stores are usually on two floors, with the ground floor mainly for food and the first floor for clothing, electronics and entertainment. Some stores that did not have the second floor have been converted to this format in recent years. Most Tesco Extra stores have a café and as of October 2009, all stores have a Tesco Tech Support Team.

Tesco Superstores

Tesco superstores are standard large supermarkets, stocking groceries and a much smaller range of non-food goods than Extra stores. The stores have always previously been branded as simply 'Tesco', but a new store in Liverpool was the first to use the format brand 'Tesco Superstore' above the door.[56]

Tesco Metro

Tesco Metro stores are sized between Tesco superstores and Tesco Express stores, with stores averaging 11000 square feet (1000 m^2). They are mainly located in city centres, the inner city and on the high streets of towns. The first Tesco Metro opened in Neston in 1980. Since then all Tesco branches with a high street format, including those that opened before the Covent Garden branch, have been rebranded from Tesco to Tesco Metro. The Tesco store in Devizes was the last store to finish rebranding, in September 2006. The store had not been renovated for over 20 years.

Tesco Metro, St Albans, England

Tesco Express

Tesco Express, Hilperton Road, Trowbridge, Wiltshire

Tesco Express stores are neighbourhood convenience stores averaging 2200 square feet (200 m^2), stocking mainly food with an emphasis on higher-margin products (due to small store size, and the necessity to maximise revenue per square foot) alongside everyday essentials. They are found in busy city centre districts, small shopping precincts in residential areas, small towns and villages and on Esso petrol station forecourts. The 1000th Tesco Express site opened in July 2009. Tesco have now started building Tesco Express stores with only 'Assisted-Service' tills, in which the customer scans all their own shopping and packs it, with the support of supervising staff when required.

One Stop

One Stop, which includes some of the smallest stores (smaller than a Tesco Express), is the only Tesco store format in the UK that does not include the word Tesco in its name. The brand, along with the original stores, formed part of the T&S Stores business but, unlike many that were converted to Tesco Express, these kept their old name. Subsequently, other stores bought by Tesco have been converted to the One Stop brand. Some have Tesco Personal Finance branded cash machines.

One Stop, Trowbridge, Wiltshire

The business has attracted some controversy, as grocery prices in these shops, often situated in less well-off areas, can be higher than nearby Tesco branded stores, highlighted in The Times 22 March 2010: "Britain's biggest supermarket uses its chain of 520 One Stop convenience stores—which many customers do not realise it owns—to charge up to 14 per cent more for goods than it does in Tesco-branded stores."

Tesco responded to the article stating "It is a separate business within the Tesco Group, with its own supply chain and distribution network. One Stop stores offer a different range to Express stores and its operating costs are different. One Stop's price strategy is to match its nearest competitor Cost Cutter and is frequently cheaper."[57] They can usually be found in smaller communities across the United Kingdom.

Tesco Homeplus

History

Tesco Homeplus is not Tesco's first non-food only venture in the UK. Until the late 1990s/early 2000s there were several non-food Tesco stores around the country including Scarborough and Yate. Although not in a warehouse style format, the stores were located on high streets and shopping centres, they did stock similar items to Homeplus stores. In both cases this was because another part of the shopping centre had a Tesco Superstore that stocked food items only.

In May 2005 Tesco announced a trial non-food only format near Manchester and Aberdeen,[58] and the first store opened in October 2005:

A further five stores opened before it stopped being a trial, and there is now a plan to open many more stores.

Current

Stores offer all of Tesco's ranges except food in warehouse-style units in retail parks. Tesco is using this format because only 20% of its customers have access to a Tesco Extra, and the company is restricted in how many of its superstores it can convert into Extras and how quickly it can do so. Large units for non-food retailing are much more readily available.

There are currently 13 Homeplus stores nationwide. The newest Homeplus store opened in Chester in July 2009.

Two more were due to open in the first half of 2009 at sites around the country. All of these were to feature the Order and Collect desk where customers can purchase and collect most items straightaway.

Store facts

As of 26 February 2011, at the end of its 2010/11 financial year, Tesco's UK store portfolio was as follows.[2]

Format	Number	Total area (m²)	Total area (sq ft)	Mean area (m²)	Mean area (sq ft)	Percentage of space	+/- Stores 2010/11
Tesco Extra	212	1,400,885	15,079,000	6,608	71,127	41.08%	▲ 22
Tesco Superstores	470	1,297,112	13,962,000	2,760	29,706	38.04%	▲ 15
Tesco Metro	186	194,632	2,095,000	1,046	11,263	5.73%	▲ 5
Tesco Express	1,285	272,392	2,932,000	212	2,282	7.99%	▲ 155
One Stop	599	74,044	797,000	142	1,530	2.04%	▲ 8
Tesco Homeplus	13	51,468	554,000	3,790	42,615	1.53%	— 0
Dobbies	28	121,053	1,303,000	4,323	46,536	3.57%	▲ 3
Total	**2,715**	**3,411,586**	**36,722,000**	**1,257**	**13,526**	**100%**	▲ 208

Other businesses

Garden centres

Tesco announced its intention to purchase Dobbies Garden Centres for £155.6 million on 8 June 2007. Dobbies operates 28 garden centres, half in Scotland and half in England.[59] The deal was confirmed as successful by the board of directors of Tesco on 17 August 2007 when the board announced that they had received 53.1% of shares (or 5,410,457 shares), which confirmed conditions set out in the offer made on 20 June 2007. Although the deal had been confirmed by Tesco the offer remained open to Dobbies shareholders until 20 August 2007.[60] Tesco raised its holding to 65% in September[61] and on 5 June 2008[62] Tesco announced that it would be compulsorily acquiring Dobbies Garden Centres plc. Dobbies continues to trade under its own brand, from its own head office in Melville, near Edinburgh.

Banking

Tesco has a banking arm called Tesco Bank, formerly a 50:50 joint venture with the Royal Bank of Scotland. Products on offer include credit cards, loans, mortgages, savings accounts and several types of insurance, including car, home, life and travel. They are promoted by leaflets in Tesco's stores and through its website. The business made a profit of £130 million for the 52 weeks to 24 February 2007, of which Tesco's share was £66 million. This move towards the financial sector diversified the Tesco brand and provides opportunities for growth outside of the retailing sector.

On 28 July 2008 Tesco announced that they were buying out the Royal Bank of Scotland's 50% stake in the company for £950 million.[63] [64] In October 2009 the name of Tesco Personal Finance was changed to Tesco Bank.

Technika

Technika is a brand name for Electronic products sold exclusively through Tesco Stores. The **Technika** range currently includes Televisions, MP3 Docking Stations, Computer Peripherals, DVD and Blu-ray Players, DAB Radios. The range is updated on a regular basis to follow market trends.

The Technika brand is managed in-house by Tesco alongside its other brands, such as Tesco and Tesco Value. Customer support is offered through the Tesco Electrical Helpline [65] or in-store through Tesco Tech Support [66].

Telecoms

Tesco operates mobile phone, home phone and broadband businesses. These are available to residential consumers in several countries and are sold via the Tesco website and through Tesco stores. Tesco has not purchased or built a telecoms network, but instead has pursued a strategy of pairing its marketing strength with the expertise of existing telecoms operators.

Mobile

In autumn 2003, Tesco Mobile was launched as a joint venture with O2. In November 2009 Tesco announced over 2 million UK customers are using this service. A similar O2 based service has since been launched in Ireland and Slovakia.

ISP

In August 2004 Tesco broadband, an ADSL-based service delivered via BT phone lines, was launched in partnership with NTL. In November 2009 Tesco announced a new partnership with Cable & Wireless, and a fresh focus on this business area.

VoIP

In January 2006, Tesco Internet Phone, a Voice over Internet Protocol, VoIP, service was launched in conjunction with Freshtel of Australia. This service was shut down in 2010.

Phone Shops

In November 2009 Tesco announced it now has 100 Phone Shops embedded within larger Tesco Extra stores, and stated an intention to open up to 500 such shops across the UK in the medium term. In April 2010 the first Tesco Phone shop opened in Slovakia.

Fuel

Tesco first started selling petrol in 1974. Tesco sells 95, 97 and 99 RON (a fuel developed by Greenergy of which Tesco is a shareholder) petrol on a retail basis from forecourts at most superstore and Express locations. Tesco have recently diversified into biofuels, offering petrol-bioethanol and diesel-biodiesel blends instead of pure petrol and diesel at their petrol stations, and now offering Greenergy 100% biodiesel at many stores in the southeast of the United Kingdom.

Tesco petrol station in St Mellons

On 28 February 2007 motorists in South East England reported that their cars were breaking down. This was due to petrol sold by Tesco and others being contaminated with silicon,[67] Tesco has been criticised with claims that they had been alerted to the problem as early as 12 February 2007. On 6 March, Tesco offered to pay for any damage caused by the faulty petrol, after printing full page apologies in many national newspapers.[68]

Tech Support

Tesco acquired a small I.T. support company called The PC Guys in 2008, and were able to launch Tesco Tech Support in December of that year.[69] Teams of Advisors were put into all Extra stores with the sole job role of answering technical questions on Tesco's range of electrical products. They also are responsible for advising customers on extended warranties, electrical returns and a range of payment plans on all electrical goods over a certain amount. Through the Cardiff Customer Service Centre in the United Kingdom, Tesco Tech Support provides UK and Ireland with technical support via telephony system on the electrical products sold in their stores.

Tesco Tech Support logo

Film Making

In 2010 Tesco started funding a small film studio intended to produce Tesco exclusive direct-to- films. The first film comes out on 6 September and is called *Paris Connections*. It is based on a popular novel by Jackie Collins, and is an investigation thriller. Jackie Collins rewrote the novel to be more appropriate to the medium of film.[70] [71] [72]

Record Label

In 2010, Tesco announced that they will be forming their own record label, with notable signings since including Mick Hucknall and Nadine Coyle. Tesco records will be exclusive products to Tesco stores.

Video-on-Demand

On 20 April 2011, Tesco acquired a 80% stake in Blinkbox from Eden Ventures and Nordic Venture Partners, it intends to use the company to boost its digital entertainment offering.[73]

Gold Exchange

In 2011, Tesco launched Tesco Gold Exchange, which is a postal gold service, offering money for gold, as well as offering clubcard points to customers via their website.[74]

Tesco Tyres

In 2011, Tesco launched tesco-tyres.com in association with Blackcircles.com, offering a choice of over 1,200 fitting partners across the UK as well as offering clubcard points with purchases.

Your Beauty Salon

In February 2011, Tesco launched Your Beauty Salon, in Tesco stores planning to open 70 over the next year, offering services like haircuts, leg waxing, manicures and eyebrow.[75]

Tesco Clubcard

Customers can collect one Clubcard points for every £1 (or one point for €1 in Ireland and Slovakia) they spend in a Tesco store, or Tesco.com, and 1 point per £2 on fuel (not in Slovakia). Customers can also collect points by paying with a Tesco Credit Card, or by using Tesco Mobile, Tesco Homephone, Tesco Broadband, selected Tesco Personal Finance products or through Clubcard partners, E.ON and Avis. Each point equates to 1p in store when redeemed or up to 3 times their value when used with clubcard deals (offers for holidays, day trips, etc.) Clubcard points (UK & IE) can also be converted to Airmiles.

Tesco have recently annonced a partniship with Virgin Atlantic whereby you can convert clubcard points into frequent flyer miles [76]

Holders receive Clubcard statements 6 times a year, which often feature extra point coupons and money-off coupons. These can be spent in-store, online or on various Clubcard deals.

Tesco was cited in a Wall Street Journal article[77] as using the intelligence from the Clubcard to thwart Wal-Mart's initiatives in the UK.

Internet operations

Tesco operates a grocery homeshopping service, as well as providing consumer goods, telecommunications and financial services online.

In May 1984, in Gateshead, England, the world's first recorded online home shopper, Mrs Jane Snowball, purchased groceries from her local Tesco store in the world's first recorded online shopping transaction from the home.[78]

Tesco has operated on the internet since 1994 and was the first retailer in the world to offer a robust home shopping service in 1996. Tesco.com was formally launched in 2000. It also has online operations in the Republic of Ireland and South Korea. Currently it's also considering entering Polish market. Grocery sales are available within delivery range of selected stores, goods being hand-picked within each store, in contrast to the warehouse model followed by Ocado. In 2003, tesco.com's CEO at the time, John Browett, received the Wharton Infosys Business Transformation Award for the innovative processes he used to support this online grocery service.

On 1 October 2007, Tesco announced that it will be selling six own-brand budget software packages for under £20 each, including office and security suites, in a partnership with software firm Formjet.[79] As Formjet is exclusive distributor for Panda Software and Ability Plus Software, packages from these companies are likely to feature.

Tesco offers an internet-based DVD rental service, which is operated by LOVEFiLM and a music download service.

International operations

Tesco's international expansion strategy has responded to the need to be sensitive to local expectations in other countries by entering into joint ventures with local partners, such as Samsung Group in South Korea (Samsung-Tesco Home plus), and Charoen Pokphand in Thailand (Tesco Lotus), appointing a very high proportion of local personnel to management positions. It also makes small acquisitions as part of its strategy for example, in its 2005/2006 financial year it made acquisitions in South Korea, one in Poland and one in Japan.[80]

In late 2004 the amount of floorspace Tesco operated outside the United Kingdom surpassed the amount it had in its home market for the first time, although the United Kingdom still accounted for more than 75% of group revenue due to lower sales per unit area outside the UK.

In September 2005 Tesco announced that it was selling its operations in Taiwan to Carrefour and purchasing Carrefour's stores in the Czech Republic and Slovakia. Both companies stated that they were concentrating their efforts in countries where they had strong market positions.

The following table shows the number of stores, total store size in area and sales for Tesco's international operations. The store numbers and floor area figures are as at 27 February 2010. This information is taken from the 2009/10 financial broker pack [81]PDF (106 KB).

Country	Entered	Stores	Area (m² (sq ft))	Mean Store Area (m² (sq ft))	+/- Stores 2010/11
China	2004	105	755,766 (8,135,000)	7,198 (77,476)	▲17
Czech Republic	1996	158	486,626 (5,238,000)	3,080 (33,152)	▲22
Hungary	1994	205	667,508 (7,185,000)	3,256 (35,049)	▲29
Republic of Ireland	1997	130	301,099 (3,241,000)	2,316 (24,931)	▲11
Japan	2003	140	46,916 (505,000)	335 (3,607)	▼2
Malaysia	2002	44	308,903 (3,325,000)	8,129 (87,500)	▲7
Poland	1992	371	756,788 (8,146,000)	2,040 (21,957)	▲35
Slovakia	1996	97	314,291 (3,383,000)	3,240 (34,876)	▲16
South Korea	1999	354	1,084,736 (11,676,000)	3,064 (32,983)	▲49
Thailand	1998	782	1,053,786 (11,342,000)	1,348 (14,504)	▲119
Turkey	2003	121	594,487 (6,399,000)	4,913 (52,884)	▲16
United Kingdom (for comparison)	1919	2,715	3,411,586 (36,722,000)	1,257 (13,526)	▲205
United States	2007	164	199,091 (2,143,000)	1,214 (13,067)	▲19
Total (not including UK)		**2,666**	**6,569,997 (70,718,000)**	**Mean: 2,465 (26,536)**	▲336
Total (including UK)		**5,381**	**9,981,583 (107,440,000)**	**Mean: 1,855 (19,970)**	▲544

China

Tesco acquired a 50% stake in the Hymall chain, from Ting Hsin in September 2004. In December 2006 it raised its stake to 90% in a £180 million deal.[82] Most of Tesco China's stores are based around Shanghai, but according to Tesco it plans to equip the business to expand more quickly and in different areas. Tesco has a large store in Weifang, Shandong province. Tesco has been increasing its own brand products into the Chinese market as well as introducing the Tesco Express format.[83]

Czech Republic

Tesco opened its first store in the Czech Republic in 1996 and now has over 84 stores, with further planned.[84] Tesco opened its first stores in the Czech Republic by buying US corporation Kmart's operations in the country and converting them into Tesco stores. Tesco is also keen to expand non-food items and has already opened petrol stations and offers personal finance services in the Czech Republic.[85] There are currently three Tesco Extra stores in the Czech Republic – one in Prague and two in Plzeň.

Tesco Express, Belehradska Street, Prague, Czech Republic

France

Tesco owned a French food retailer called Catteau between 1992 and 1997, which operated a chain of stores in NE France under the Cedico, Hyper Cedico and Cedimarche banners. Tesco also operated a "Vin Plus" outlet in Calais, selling wine, beer and spirits, which closed on 30 August 2010.[86]

Hungary

Tesco launched in Hungary in 1994 after purchasing a small local supermarket group trading as S-Market and based in the North West of Hungary. It opened its first hypermarket in Hungary at the Polus Centre in Budapest in 1996. Tesco operates through more than 200 stores in Hungary with further openings planned.[84] Tesco offers its value, standard, healthy living and finest range in its stores. Tesco Hungary also offers a clothing line and personal finance services.[87] In August 2010 opened the first Tesco Extra in Budapest; its name is Tesco Extra Fogarasi and it is located in Zugló, Budapest.

Tesco store at Kőszeg, Hungary with Statue of Fire Salamander

Republic of Ireland

Tesco operated in the Irish grocery market in the early eighties, however sold its operations there in March 1986.[88]

Tesco re-entered the Irish market in 1997 after the purchase of Power Supermarkets Ltd.

The country's newspaper of record the Irish Times in April 2011 said that "Increasingly, Ireland is being viewed as a provincial backwater by the parent company – albeit a very profitable little backwater – and all the strategic decisions are being taken in the UK.[89]

It now operates from 101 stores across Ireland. Like Tesco stores in the UK, these offer a home delivery shopping service available to 80% of the Irish population as well as petrol, mobile telephone, personal finance, flower delivery service and a weight-loss programme.[90] Also available is Tesco's loyalty programme, the Clubcard.

Tesco is now the grocery market leader in the Republic of Ireland, with a reported November 2005 share of 26.3%.[91] Tesco Ireland also claims to be the largest purchaser of Irish food with an estimated €1.5 billion annually. Tesco Ireland operates a small number of Tesco Extra hypermarkets in the Republic of Ireland, with Clarehall Extra on the Malahide Road being the first to open in 2006. Tesco's largest hypermarket store in Europe, with a floorspace of 18,500 m², opened in Naas in Co Kildare in November 2010.[92]

India

Tesco has had a limited presence in India with a service centre in Bangalore, and outsourcing.[93] However, in 2008 Tesco announced their intention to invest an initial £60m ($115m) to open a wholesale cash-and-carry business based in Mumbai with the assistance of the Tata Group.[94]

Isle of Man

Tesco opened a large purpose built store along with a car park in Douglas on the Isle of Man in 2002, There were plans to rebuild the store to almost double the floor space, including a mezzanine level and to replace the open car parking area with a two storey car park, However these plans fell through in 2010. Tesco also runs a home delivery service across the island. There are currently (as of 2011) plans to open a second store in Castletown, Isle of Man.

Japan

Tesco Japan first began operations in 2003. It was brought about by a buy-out of C Two stores for £139 million in July 2003 and later Fre'c in April 2004.[95] Tesco has adopted an approach that focuses on small corner shops that operate similarly to its Express format, rather than opening hypermarkets. It has also launched its range of software in Japan.[84] In August 2011, Tesco announced that they would be selling off their Japanese stores after revealing that only half of the stores in the Greater Tokyo Area were making a profit.[96]

Malaysia

Tesco opened its first store in Malaysia in May 2002 with the opening of its first hypermarket in Puchong, Selangor. Tesco Malaysia currently operates 45 Tesco and Tesco Extra stores. Selangor has 12 stores, Perak seven stores, Johor six stores, Kuala Lumpur and Penang five stores each, Kedah four stores, Negeri Sembilan three stores, Melaka two stores, and Kelantan one store. Tesco has partnered with local conglomerate Sime Darby Berhad, which holds 30% of the shares.[97]

A Tesco outlet in Bandar Bukit Tinggi, Klang, Selangor, Malaysia.

Tesco also acquired the Malaysian operation of the wholesaler Makro, which was rebranded Tesco Extra and provides products for local retailers. As of 2011, Tesco has relaunched the Tesco Extra brand in five of its stores in the Klang Valley. The new Tesco Extra brand will now offer the widest choice in the food, clothing, home and electronics ranges. A variety of complementary services such as a pharmacy, an optician and a Tesco phone shop will also be incorporated into these Tesco Extra stores.

Tesco Malaysia offers a value range, its own branded range, electronic goods, the loyalty clubcard and clothing. Tesco Malaysia's clubcard introduced Green ClubCard Points in 2007 making Tesco Malaysia the first Tesco international business to introduce the scheme (Green ClubCard Points).[98]

Poland

Tesco entered the Polish market in 1992. It currently operates from 334 stores.[84] Tesco Poland offers the value, healthy living and own branded line of products as well as regional produce, petrol, personal finance services and on-line photo processing. In August 2008 Tesco opened the first Extra store in Poland located in Częstochowa. Currently there are five Tesco Extra stores in Poland.

A Tesco Hypermarket in Prokocim, Poland

Slovakia

Tesco Slovakia opened in 1996 as part of Tesco's international expansion aims. It now operates from 97 stores and 16 service stations.[99] Tesco Slovakia has recently put great emphasis on organic products. However, Tesco Slovakia caused controversy amongst the Slovak government when it was found to have come foul of food safety laws in 2006.[100] In April 2010 the first Tesco Extra in Central Europe opened in Bratislava – Petrzalka, Slovakia as part of a pilot project for Tesco in the region, including the first self-service cash flow in Central Europe. There are currently four Tesco Extra stores in Slovakia – three in Bratislava and one in Zvolen.[101]

Tesco, Kamenné námestie, Bratislava, Slovakia

South Korea

Tesco launched its South Korean operations in 1999 and partnered with Samsung, currently Tesco holds 94% of the shares in the venture.[102] It operates both hypermarkets and its express format as well as a home delivery shopping service. It is the second largest retailer in South Korea, just behind Shinsegae Group.[102]

On 14 May 2008, Tesco agreed to purchase 36 hypermarkets with a combination of food and non-food products from E-Land for $1.9 billion (£976 million) in its biggest single acquisition, making Tesco the second largest in the country. The majority of the E-Land stores formerly belonged to French retailer Carrefour before 2006 and most of the stores will be converted to Tesco Homeplus outlets. Tesco's South Korean discount store chain, Home Plus, currently has 66 outlets.[103] [104]

Taiwan

Tesco entered Taiwan in September 2004. In September 2005, Tesco sold its stores in Taiwan to Carrefour.

Thailand

Tesco entered Thailand in 1998 and operates through 380 stores as part of a joint venture with Charoen Pokphand and named the operation Tesco Lotus. This partnership was dissolved in 2003 when Charoen Pokphand sold its shares to Tesco. Tesco Lotus sells a diverse range of products from value food products to electronics to personal finance services. The company is keen to promote its green values and has partnered with the UNEP. Tesco Lotus claims to serve 20 million customers every month and that 97% of its goods are sourced from Thailand.[105]

Tesco-Lotus superstore in Sakon Nakon, Thailand. Shoppers have to pass through a checkpoint before entering the premises.

Turkey

Tesco entered Turkey in 2003 and uses the trading name "Kipa". Tesco remains focused on building infrastructure in Turkey to complete its expansion plans and has already introduced the Tesco Express format into Turkey. There are plans to increase the rate of expansion as basic infrastructure is built.[84] The first Tesco Extra in Turkey opened in Izmir, Turkey, Tesco Kipa Extra Balçova in September 2010.

United States of America

Tesco moved into the United States market in 2007, opening a chain of grocery convenience stores, named Fresh & Easy, on the West Coast (Arizona, California and Nevada).[106] The company established its U.S. headquarters in El Segundo, California, and the first store opened in Hemet, California in November 2007, with 100 more expected in the first year; they planned to open a new one every two-and-a-half days in the United States.

A Fresh & Easy store in Summerlin, Nevada, United States

Although the planned rate of expansion was not maintained, largely because of the recession, Fresh & Easy now (August 2011) operates 182 stores in Arizona, California and Nevada.

Financial performance

Tesco is listed on the London Stock Exchange under the symbol **TSCO**. It also has a secondary listing on the Irish Stock Exchange with the name **TESCO PLC**.

All figures below are for the Tesco's financial years, which run for 52 or 53 week periods to late February. Up to 27 February 2007 period end the numbers include non-UK and Ireland results for the year ended on 31 December 2006 in the accounting year. The figures in the table below include 52 weeks/12 months of turnover for both sides of the business as this provides the best comparative.

52/3 weeks ended	Turnover (£m)	Profit before tax (£m)	Profit for year (£m)	Basic earnings per share (p)
26 February 2011	67,573	3,535	2,671	33.10
27 February 2010	62,537	3,176	2,336	31.66
28 February 2009	54,300	3,128	2,166	28.92
23 February 2008	47,298	2,803	2,130	26.95
24 February 2007	46,600	2,653	1,899	22.36
25 February 2006	38,300	2,210	1,576	19.70
26 February 2005	33,974	1,962	1,366	17.44
28 February 2004	30,814	1,600	1,100	15.05
22 February 2003	26,337	1,361	946	13.54
23 February 2002	23,653	1,201	830	12.05
24 February 2001	20,988	1,054	767	11.29
26 February 2000	18,796	933	674	10.07
27 February 1999	17,158	842	606	9.14
28 February 1998	16,452	760	532	8.12

As of its 2006 year end Tesco was the fourth largest retailer in the world behind Wal-Mart, Carrefour and Home Depot. Tesco moved ahead of Home Depot during 2007, following the sale of Home Depot's professional supply division and a decline in the value of the U.S. dollar against the British Pound. METRO was only just behind and might move ahead again if the euro strengthens against the pound, but METRO's sales include many billions of wholesale turnover, and its retail turnover is much less than Tesco's.

At 24 February 2007 Tesco operated 1,988 stores in the UK with 27700000 square feet (m^2) of floorspace and 1,275 outside the UK with 4040000 square feet (m^2) of floorspace.

Despite being in a recession, Tesco made record profits for a British retailer in the year to February 2010, during which its underlying pre-tax profits increased by 10.1% to £3.4 billion. Tesco now plans to create 16,000 new jobs, of which 9,000 will be in the UK.[107] In 2011 the retailer reported its poorest six-monthly UK sales figures for 20 years, as a result of consumers' reduced non-food spending.[108]

UK market share

According to TNS Worldpanel, Tesco's share of the UK grocery market in the 12 weeks to 27 December 2009 was 30.5%, up 0.1% on 12 weeks to 27 December 2008 largely in line with the increase of the other four largest supermarkets.[109] These increases follow a decline in the market share of discount chains Lidl, Aldi and Netto in the same period, collectively of 0.1%.

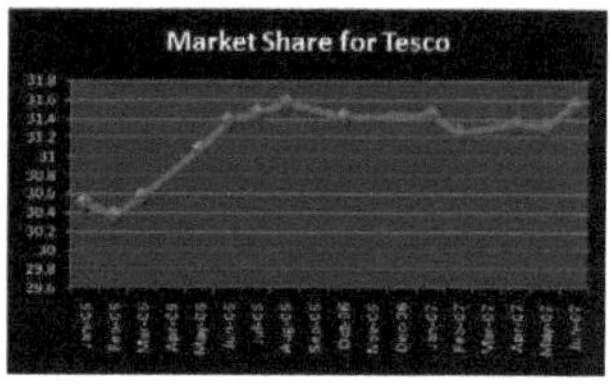

Graph Showing UK Market Share of Tesco

Supermarket	Market Share December 2009	+/- from December 2008
Tesco	30.5%	▲ 0.1%
Asda	16.9%	▲ 0.1%
Sainsbury's	16.3%	▲ 0.2%
Morrisons	12.3%	▲ 0.5%

Tesco litigation

As with any large corporation, Tesco is involved in litigation, usually from claims of personal injury from customers, claims of unfair dismissal from staff, and other commercial matters. Two notable cases were *Ward v Tesco Stores Ltd*, which set a precedent in so called 'trip or slip' injury claims against retailers; and *Tesco Supermarkets Ltd v Nattrass*, which reached the House of Lords, and became a leading case regarding the corporate liability of businesses for failures of their store managers (in a case of misleading advertising).

Criticism

Tesco have been criticised for aggressively pursuing critics of the company in Thailand. Writer and former MP Jit Siratranont is facing up to two years in jail and a £16.4 million libel damages claim for saying that Tesco was expanding aggressively at the expense of small local retailers. Tesco served him with writs for criminal defamation and civil libel.[110]

Criticism of Tesco includes allegations of stifling competition due to its undeveloped "land bank",[111] and breaching planning laws.[112]

In December 2006 *The Grocer* magazine published a study that named Tesco as having the slowest checkouts of the six major supermarkets. Somerfield had the shortest queues with an average wait of 4 min 23 seconds. In order of least time spent at the checkout, the other major supermarkets were Waitrose, Sainsbury's, Asda, Morrisons.[113]

The Grocer also named Asda as the cheapest UK supermarket (based on 33 items). Tesco was second and Sainsbury's and Morrisons joint third.[113] Tesco price check tends to differ saying out of 7134 (compared to Asda) products, (Survey carried out between 9 July 2007 and 11 July 2007) Tesco is cheaper: 1835 (compared to 1251 the previous week), Tesco is more expensive: 975 (compared to 984 the previous week) and Tesco is the same price: 4324 (compared to 4996 the previous week).[114]

In 2009, Tesco continued to advertise on Fox News's controversial Glenn Beck Show even after 60 major US companies, and a number of UK companies, pulled their advertisements. The cancellations were largely a consequence of Beck's accusation that Barack Obama was racist and had a "deep-seated hatred for white people".[115]

Corporate tax structure

In May 2007, it was revealed that Tesco had moved the head office of its online operations to the tax haven of Switzerland. This allows it to sell CDs, DVDs and electronic games through its web site without charging VAT.[116] The operation had previously been run from the tax haven of Jersey, but had been closed by authorities who feared damage to the Island's reputation.[116] In June 2008, the government announced that it was closing a tax loophole being used by Tesco.[117] The scheme, identified by British magazine *Private Eye*, utilises offshore holding companies in Luxembourg and partnership agreements to avoid a corporation tax liability of up to £50 million a year.[117] Another scheme previously identified by Private Eye involved depositing £1 billion in a Swiss partnership, and then loaning out that money to overseas Tesco stores, so that profit can be transferred indirectly through interest payments. This scheme is still in operation and is estimated to be costing the UK exchequer up to £20 million a year in corporation tax.[117] Tax expert Richard Murphy has provided an analysis of this avoidance structure.[118]

Legal action against *The Guardian* newspaper

In February 2008 a six-month investigation by *The Guardian* claimed that Tesco had developed a complex taxation structure involving offshore bank accounts in the tax haven of the Cayman Islands.[119] Tesco was in the process of selling its UK stores, worth an estimated £6 billion, to Cayman Island-based companies set up by Tesco. These companies then lease the stores back to Tesco. At the time, *The Guardian* claimed that this arrangement would enable Tesco to avoid an estimated £1 billion tax on profits from the property sales, and also avoid paying tax on continuing operation of the stores, as the rate of corporation tax in the Cayman Islands is zero. Tesco defended its tax arrangements, saying it had a legal duty to its shareholders to organise its affairs in a tax-efficient manner and pointing out that the company already pays significant amounts of tax, including VAT, excise duty and fuel duty on behalf of its customers, PAYE and national insurance contributions on behalf of its employees and corporation tax on its profits.

Following these revelations, several MPs called for an inquiry into Tesco's tax avoidance schemes.[120]

Tesco issued a writ for libel against *The Guardian* five weeks later. Tesco denied that it had avoided paying £1 billion corporation tax but for legal reasons refused to answer further questions or to clarify the purpose of the tax structure they had created. Further investigations by *The Guardian* discovered that the tax structures were aimed at avoiding Stamp Duty Land Tax, and not corporation tax as originally thought. SDLT is levelled at 4%, and corporate tax at around 30%, so the figure of £1 billion tax avoided by Tesco has been revised to an estimated £90–£100 million.[121] According to *the Guardian*, "Tesco has been involved in a game of cat and mouse with HM Revenue & Customs since 2003. On three occasions when the government has closed a loophole to prevent avoidance, Tesco has taken advantage of ingenious schemes to get around it. Tesco still has 36 stores wrapped up in UK limited partnerships—with Cayman Islands registered partners—which were established in 2006 before the latest loophole was closed."[121]

In September 2008, Tesco dropped the legal action after *The Guardian* apologised for its story, acknowledging "These damaging allegations were unfounded and should not have been published. All profits generated by this sale and leaseback arrangement were earned by UK tax-resident companies and have been or will be included in Tesco's UK tax returns. The use of Cayman Island companies in the scheme was for legitimate stamp duty savings purposes."[122] Costs were paid by the newspaper with undisclosed damages being paid to charity.[122]

The Guardian's parent company, Guardian Media Group, used a similar offshore arrangement to avoid tax during its takeover of Emap.[123]

Opposition to expansion

Tesco's expansion has not been without criticism and, in some cases, active opposition.

- A 2006 article on the thisismoney website quoted various criticisms of the company's expansion and dominance of the UK supermarket scene.[124]
- In 2007 residents in the Hall Green area of Birmingham made their opposition strongly evident when Tesco announced plans to open a store on the Robin Hood Lane, right next to a busy island in the suburb. The Local Liberal Democrat councillor actively opposed the store on the grounds that it would affect local businesses and consumer choice. Tesco received planning permission to open the store in April 2008 and residents made plans to boycott the store and continue to use local outlets. Since opening Tesco's deliveries have caused major traffic problems in the area and Birmingham City Council have announced plans to put a loading ban into force between 7:00 am and 10:00 am to relieve congestion.[125]
- In 2008, opposition to a proposed expansion in Coventry was reported in the *Coventry Telegraph*.[126]
- In 2007 residents of Cambridge organised a campaign, "No Mill Road Tesco" that aimed to prevent Tesco from opening a store on Cambridge's Mill Road. This street runs just to the south of the centre of Cambridge which is a lively, cosmopolitan area home to many independent shops, delicatessens, cafes, and curry houses.[127] In March 2008, Tesco's application was refused by the council. Tesco appealed, but lost the appeal in November 2008.[128] The planning inspector dismissed the appeal on the grounds of highway safety.[129] In mid-2008, while waiting for the appeal to be heard, Tesco applied for planning permission for an air conditioning and refrigeration plant, which was also rejected by the council.[130] Tesco eventually opened a restricted size store with no external refrigeration plant or alcohol license in August 2009.[131]
- In March 2007 residents in Bournville, Birmingham fought to maintain the historic alcohol free status of the area, in winning a court battle with Tesco, to prevent it selling alcohol in its local outlet. No shops are permitted to sell alcohol in the area and there are no pubs, bars or fast-food outlets in Bournville.[132]
- Plans for a large Tesco store in St Albans, Hertfordshire, attracted widespread local opposition. This led to the formation of the "Stop St Albans Tesco Group". In June 2008, St Albans Council refused planning permission for the proposed store.[133]
- In 2008, Tesco faced opposition to plans to build a new store in Tonypandy, South Wales. Local business protested against the plans and Tesco was later refused planning permission by the local council.[134] Tesco later said they would appeal to the Welsh assembly, However the result of this has not emerged.
- In 2009, Tesco received planning permission to build a Tesco Express store on Hope Street, Liverpool despite there being a total of eight other Tesco stores (In Express, Metro and Superstore formats) within less than or equivalent to a mile from its proposed location.[135] This initiated a campaign in the local area[136] and a large Facebook group movement to prevent the construction going ahead.[137] Tesco withdrew the plans on 3 September 2009 due to widespread opposition and condemnation from local people.
- In April 2011, longstanding opposition to a Tesco Express store in Cheltenham Road, Stokes Croft, Bristol, evolved into a violent clash between opponents and police. The recently-opened storefront was heavily damaged, and police reported the seizure of petrol bombs.[138] Opponents have suggested that the store would damage small shops and harm the character of the area.[139]
- Sheringham in Norfolk, has been fighting Tesco for over 13 years. Since 1997 Tesco have had four planning permissions rejected and an appeal thrown out by the planning inspector[140] and by members of the public.[141]

Alumni

In recent years Tesco alumni have had increasing influence on other large listed businesses. Former Tesco senior staffers include:

- CEO Talk Talk plc : Dido Harding[142]
- CEO Halfords plc : David Wild[143]
- CEO Figleaves.com : Julia Reynolds[144]
- CEO Dixons plc : John Browett[145]
- Deputy Chair Carphone Warehouse: John Gildersleeve[146]
- CEO Carrefour France : James McCann[147]
- CEO Greggs plc : Ken McMeikan[148]
- CEO Domino's Pizza UK & IRL : Lance Batchelor[149]

Further reading

- MacLaurin, Sir Ian (1999). *Tiger by the Tail: A Life in Business from Tesco to Test Cricket.* London: Pan Books. ISBN 0330373714.
- Simms, Andrew (2007). *Tescopoly: how one shop came out on top and why it matters.* London: Constable. ISBN 1845295110.
- Humby, Clive; Hunt, Terry & Phillips, Tim (2006). *Scoring points : how Tesco continues to win customer loyalty.* London & Philadelphia: Kogan Page. ISBN 9780749447526.
- Nash, Bethany (2006). *Fair-Trade and the growth of ethical consumerism within the mainstream : an investigation into the Tesco consumer.* Leeds: University of Leeds. ISBN 75272130.

Gallery

Tesco in Burscough, Lancashire

Tesco in Baldock, Hertfordshire

Tesco in Swindon, Wiltshire

Tesco Store at Národní Třída metro station, Prague

Czech Tesco HQ, Prague

Tesco Brno

Tesco Mikulov, Czech Republic

Tesco Express near Emirates Stadium

Tesco Store in Northallerton, North Yorkshire

Tesco Hypermarket in Zličín, Czech Republic

Tesco Metro in Oxford

See also

- List of supermarket chains in the United Kingdom
- Tescopoly
- TNS Worldpanel
- Tesco Town

References

[1] http://www.londonstockexchange.com/exchange/prices-and-news/stocks/prices-search/stock-prices-search.html?nameCode=TSCO

[2] "Preliminary Results 2011" (http://www.tescoplc.com/plc/ir/pres_results/results/r2011/2011-04-19/2011-04-19.pdf). 19 April 2011. . Retrieved 19 April 2011.

[3] "2010 Annual Report & Financial Statements" (http://ar2010.tescoplc.com/~/media/Files/T/Tesco-Annual-Report-2009/Attachments/pdf/tesco-annualreport.pdf). Tesco. . Retrieved 20 January 2011.

[4] http://www.tesco.com

[5] " Tesco.com Terms & Conditions (http://www.tesco.com/termsandconditions/termsconditionsGeneral.htm)." Tesco. Retrieved on 20 June 2010. "The Tesco.com Site is owned and operated by Tesco Stores Limited a company registered in England and Wales whose registered office is at Tesco House, Cheshunt, Waltham Cross, Hertfordshire, EN8 9SL."

[6] Potter, Mark (16 February 2011). "Tesco to outpace growth at global rivals – study" (http://www.reuters.com/article/2011/02/17/tesco-igd-idUSLDE71F1LR20110217). Reuters. . Retrieved 25 February 2011.

[7] "Global Powers of Retailing 2010" (http://www.deloitte.com/assets/Dcom-Australia/Local Assets/Documents/news-research/Press releases/Global Powers of Retailing/Global_Powers_of_Retailing_2010_report.pdf). Deloitte. . Retrieved 18 August 2010.

[8] "International" (http://www.tescoplc.com/plc/about_us/strategy/international/). Tesco plc. . Retrieved 1 September 2010.

[9] "Interactive map" (http://www.tescoplc.com/plc/about_us/map/). Tesco plc. . Retrieved 1 September 2010.

[10] Finch, Julia (2 February 2010). "Tesco opens its first zero carbon store" (http://www.guardian.co.uk/business/2010/feb/02/tesco-carbon-neutral-green-building). *The Guardian* (UK). . Retrieved 1 September 2010.

[11] "Our history" (http://www.tescoplc.com/about-tesco/our-history/). Tesco plc. . Retrieved 10 September 2010.

[12] Sir John Edward Cohen at Oxford Dictionary of National Biography (http://www.oxforddnb.com/view/article/30949)

[13] "Tesco DVD Rental" (http://www.tescodvdrental.com/visitor/home.html). Tesco. . Retrieved 27 July 2006.

[14] "Tesco: Our History" (http://www.tescocorporate.com/plc/about_us/tesco_story). *www.tescocorporate.com*. Tesco plc. . Retrieved 27 March 2007.

[15] "Tesco Plc: Overview" (http://www.corporatewatch.org.uk/?lid=252). . Retrieved 17 August 2007.

[16] "TESCO in 2003" (http://www.icmr.icfai.org/casestudies/catalogue/Business Strategy3/BSTA085.htm). ICFAI. . Retrieved 27 March 2007.

[17] O'Grady, Sean (16 December 2001). "Shirley Porter: Rich, flashy and corrupt with it. She's nothing like a Dame" (http://www.independent.co.uk/news/people/profiles/shirley-porter-rich-flashy-and-corrupt-with-it-shes-nothing-like-a-dame-620265.html). *The Independent* (London). . Retrieved 13 December 2009.

[18] Harvey, Oliver (9 May 2007). "TESCO kills of Well St Market" (http://www.thesun.co.uk/sol/homepage/news/25532/Tesco-kills-off-Well-St-market.html). *The Sun* (London). . Retrieved 13 December 2009.

[19] Pagano, Margareta (16 May 1987). "Shocked Hillards attacks greed of Prudential: Hartley disgusted with institutions after Tesco wins takeover battle". *The Guardian* (the sun).

[20] "Tesco accused of 'near monopoly'" (http://news.bbc.co.uk/1/hi/scotland/4618972.stm). BBC. 17 January 2006. .

[21] "Inverness: Tescotown" (http://www.tescotown.co.uk/). . Retrieved 13 March 2006.

[22] Lorna Martin (1 January 2006). "The supermarket that ate a town" (http://observer.guardian.co.uk/uk_news/story/0,6903,1676049,00.html). *The Observer* (UK). .

[23] Walker, Gaelle (11 November 2006). "Online failing todeliver". *The Grocer* (William Reed Publications): p. 6.

[24] "Business as usual for Sir Terry after 10 years in charge". *Birmingham Post* (Midland Independent Newspapers): p. 24. 22 February 2007.

[25] Buckley, Neil (22 November 1995). "People: Leahy rings Tesco's tills". *Financial Times*: p. 40.

[26] Cunningham, Sarah (22 March 1997). "Tesco pays £630m for ABF's Irish business". *The Times* (Times Newspapers).

[27] "Tesco's Irish move approved". *Financial Times*. 7 May 1997.

[28] Savill, Richard (12 June 2001). "Tesco bomb blackmailer is jailed for 16 years" (http://www.telegraph.co.uk/news/uknews/1308835/Tesco-bomb-blackmailer-is-jailed-for-16-years.html). *The Daily Telegraph* (UK). . Retrieved 12 April 2011.

[29] Helft, Miguel (2001). "Tesco buys stake in GroceryWorks" (http://findarticles.com/p/articles/mi_m0HWW/is_27_4/ai_76964759). *The Industry Standard*. .

[30] "Convenience boost for Tesco – Tesco PLC acquires One Stop, Day and Nite convenience stores from T. and S. Stores PLC" (http://www.findarticles.com/p/articles/mi_m0DQA/is_2002_Nov_7/ai_94447960). Eurofood. 7 November 2002. .

[31] "Tesco buys Japanese retailer" (http://news.bbc.co.uk/1/hi/business/2977358.stm). BBC News. 10 June 2003. .

[32] "Tesco to buy controlling stake in Kipa" (http://goliath.ecnext.com/coms2/gi_0199-2925842/Turkey-close-World-News-Tesco.html). .

[33] "Tesco snaps up Cullens and Europa" (http://news.bbc.co.uk/1/hi/business/3419635.stm). BBC News. 22 January 2004. .

[34] Tran, Mark (26 September 2005). "Tesco buys 21 Morrisons Petrol Filling Stations" (http://www.guardian.co.uk/business/2005/sep/26/supermarkets.tesco). *The Guardian* (London). .

[35] "Tesco in Poland Bid" (http://www.liverpooldailypost.co.uk/business/business-local/2006/07/10/tesco-in-poland-bid-64375-17354058/). .

[36] "Tesco feels the squeeze with its worst figures in 20 years as shoppers switch to budget rivals" (http://www.dailymail.co.uk/news/article-2045430/Tesco-UK-sales-worst-20-years-shoppers-tighten-belts.html?ITO=1490). *Daily Mail*. 6 October 2011. . Retrieved 6 October 2011.

[37] Hannah Liptrot (3 June 2005). "Tesco: Supermarket superpower" (http://news.bbc.co.uk/1/hi/business/4605115.stm). BBC. .

[38] Tesco's still pushing discount products to fight back competitors (http://news.idealo.co.uk/news/4521/tescos-still-pushing-discount-products-to-fight-back-competitors.html) Household & Leisure, 15 January 2009

[39] "Tesco's 'Steering Wheel' Strategy" (http://www.icmrindia.org/casestudies/catalogue/Business Strategy/Tesco's 'Steering Wheel' Strategy.htm). ICMR. 2005. .

[40] "Human Resources – Tesco Careers" (http://www.tesco-graduates.com/home/programmes/office/personnel). .

[41] "Tesco – CIO 100 Company – CIO UK Magazine" (http://www.cio.co.uk/cio100/tesco/1699/). Cio.co.uk. . Retrieved 16 April 2010.

[42] "Tesco storm the US with Box offering – In-Depth – CIO UK Magazine" (http://www.cio.co.uk/article/583/tesco-storm-the-us-with-box-offering/?otc=44). Cio.co.uk. . Retrieved 16 April 2010.

[43] "Technology helps Tesco beat crunch" (http://www.cio.co.uk/news/3204/technology-helps-tesco-beat-crunch/?otc=44). Cio.co.uk. . Retrieved 16 April 2010.

[44] Ian MacKinnon and David Leigh (8 April 2008). "Tesco sues critic of its expansion in Thailand for £16.4m damages" (http://www.guardian.co.uk/business/2008/apr/08/tesco.supermarkets). *The Guardian* (UK). .

[45] "PDF of The 2006 Giving List" (http://image.guardian.co.uk/sys-files/Society/documents/2006/11/08/ftse100.pdf) (PDF). *The Guardian* (London). . Retrieved 18 March 2007.

[46] "FTSE 100 companies reduce charity giving" (http://www.thirdsector.co.uk/news/Article/615827/ftse-100-companies-reduce-charity-giving/). Third Sector. 8 November 2006. .

[47] "The Work Foundation praises Tesco" (http://web.archive.org/web/20071005054452/http://www.theworkfoundation.com/aboutus/media/pressreleases/tescocorporatesocialresponsibility.aspx). The Work Foundation. Archived from the original (http://www.theworkfoundation.com/aboutus/media/pressreleases/tescocorporatesocialresponsibility.aspx) on 5 October 2007. . Retrieved 18 March 2007.

[48] "Intelligent Giving criticises Tesco's staff giving policies" (http://www.intelligentgiving.com/articles/features/get_the_boss_to_help_your_charity). Intelligent Giving. . Retrieved 18 March 2007.

[49] "BITC – Tesco Computers for Schools" (http://web.archive.org/web/20051104084131/http://www.bitc.org.uk/resources/case_studies/crmtescocfs.html). Archived from the original (http://www.bitc.org.uk/resources/case_studies/crmtescocfs.html) on 4 November 2005. . Retrieved 19 January 2006.

[50] O'Halloran, Marie (25 July 2009). "Unicef accuses Tesco of misusing charity slogan" (http://www.irishtimes.com/newspaper/ireland/2009/0725/1224251305112.html). *The Irish Times*. .

[51] McBride, Louise (26 July 2009). "Tesco in clash with Unicef" (http://www.independent.ie/business/irish/tesco-in-clash-with-unicef-1841427.html). *The Irish Independent*. .

[52] "Tesco gets suppliers collaborating on carbon emissions online" (http://retailtimes.co.uk/tesco-gets-suppliers-collaborating-on-carbon-emissions-online/). Retailtimes.co.uk. 2011-06-03. . Retrieved 2011-08-28.

[53] "Banned pesticides detected on vegetables in Tesco and other supermarkets in China" (http://www.greenpeace.org/eastasia/news/blog/banned-pesticides-detected-on-vegetables-in-t/blog/36666/). Greenpeace East Asia. 2011-09-08. . Retrieved 2011-09-30.

[54] Malkin, Bonnie (14 November 2007). "Spice Girls go shopping at Tesco" (http://www.telegraph.co.uk/news/uknews/1569317/Spice-Girls-go-shopping-at-Tesco.html). London: The Telegraph. . Retrieved 19 September 2010.

[55] "AEW Architects" (http://www.aewarchitects.com/whatwedo/foodretail/Tesco Walkden). AEW Architects. . Retrieved 16 August 2011.

[56] "First store to be branded 'Tesco Superstore'" (http://www.retail-week.com/Stores/2009/02/tesco_launches_superstore_format_at_liverpool_one_piling_pressure_on_nonfood_ri.html). Retail Week. .

[57] Leroux, Marcus (22 March 2010). "Tesco's secret chain charges customers more – Times Online" (http://business.timesonline.co.uk/tol/business/industry_sectors/retailing/article7070471.ece). *The Times* (UK). . Retrieved 11 November 2010.

[58] "Tesco announces non-food store trials" (http://www.tescocorporate.com/page.aspx?pointerid=DEC14C7BA76D451F8D81DBC4A14DE89C). . Retrieved 13 March 2006.

[59] Jordan, Dearbail (8 June 2008). "Tesco swoops on Dobbies Garden Centres" (http://business.timesonline.co.uk/tol/business/industry_sectors/retailing/article1902736.ece). *The Times* (London: Times Newspapers). . Retrieved 8 June 2007.

[60] "Offer Declared Unconditional in all Respects" (http://www.dobbies.com/tescoannoucement.aspx). *Dobbies* (Dobbies): p. 1. 17 August 2007. . Retrieved 19 August 2007.

[61] "Tesco sees UK sales growth slow" (http://news.bbc.co.uk/1/hi/business/7023306.stm). *Business* (BBC NEWS). 2 October 2007. . Retrieved 6 January 2010.

[62] "Compulsory acquisition of outstanding Dobbies Shares" (http://www.dobbies.com/compulsoryacquisitionofoutstandingdobbiesshrs.aspx). Dobbies. .

[63] "Tesco is taking full control of Tesco Personal Finance, after paying Royal Bank of Scotland £950 million for its 50% stake in the division" (http://news.sky.com/skynews/Home/Business/Tesco-Pays-Royal-Bank-Of-Scotland-950m-In-New-Personal-Finance-Deal/Article/200807415058484?lpos=Business_2&lid=ARTICLE_15058484_Tesco+Pays+Royal+Bank+Of+Scotland+%C2%A3950m+In+New+Personal+Finance+Deal). Sky News. 28 July 2008. . Retrieved 28 July 2008.

[64] Thelwell, Emma; Butterworth, Myra (28 July 2008). "Tesco eyes mortgages and current accounts in plan to take on UK's high street banks" (http://www.telegraph.co.uk/money/main.jhtml?view=DETAILS&grid=A1YourView&xml=/money/2008/07/28/bcntesco228.xml). London: The Telegraph. . Retrieved 28 July 2008.

[65] http://direct.tesco.com/help/returns.aspx

[66] http://www.tescotechsupport.com/

[67] "Silicon found to be rogue element in petrol" (http://news.independent.co.uk/uk/transport/article2323460.ece). London: Barrie Clement, The Independent. 3 March 2007. .

[68] "Retailers offer car repair refund" (http://news.bbc.co.uk/1/hi/uk/6421783.stm). BBC News. 6 March 2007. .

[69] "Tesco Tech Support" (http://tescotechsupport.com/). . Retrieved 1 July 2009.

[70] UK supermarket chain Tesco gets into the filmmaking biz (http://www.engadget.com/2010/08/11/uk-supermarket-chain-tesco-gets-into-the-filmmaking-biz/)

[71] "Tesco signs film-making joint venture" (http://news.bbc.co.uk/1/hi/8478414.stm). BBC News. 25 January 2010. . Retrieved 19 April 2011.

[72] Vanessa Thorpe, Arts and Media Correspondent (24 January 2010). "Tesco sets up film studio to adapt hit novels" (http://www.guardian.co.uk/business/2010/jan/24/tesco-film-movies-novels-adaptations). *The Guardian* (UK). . Retrieved 19 April 2011.

[73] "Tesco takes majority stake in Blinkbox" (http://www.digitalspy.co.uk/broadcasting/news/a315585/tesco-takes-majority-stake-in-blinkbox.html). Digital Spy. . Retrieved 20 April 2011.

[74] "Cash for Gold | Sell Your Unwanted Gold with Tesco Gold Exchange" (https://www.tescogoldexchange.com/). Tescogoldexchange.com. 13 April 2011. . Retrieved 19 April 2011.

[75] Daily Mail3 February 2011, 9:10 am (3 February 2011). "Tesco to try its hand at manicures and hair-dos" (http://www.thisismoney.co.uk/news/article.html?in_article_id=522205&in_page_id=2). This is Money. . Retrieved 19 April 2011.

[76] Virgin Atlantic |url=http://www.virgin-atlantic.com/en/gb/frequentflyer/fcpartners/shopping/tescoclubcard.jsp

|title=Tesco and Virgin Points

[77] Cecilie Rohwedder (6 June 2006). "Retailer in Britain Uses 'Clubcard' to Thwart Wal-Mart" (http://web.archive.org/web/20071001012148/http://pqasb.pqarchiver.com/wsj/access/1049173581.html?dids=1049173581:1049173581&FMT=ABS&FMTS=ABS:FT&date=Jun+6,+2006&author=Cecilie+Rohwedder&pub=Ms.+Fiala+Changes+Detergent;+Wall+Street+Journal&edition=Eastern+edition&startpage=A.1&type=91_1996&desc=Stores+of+Knowledge:+No.1+Retailer+in+Britain+Uses+'Clubcard'+to+Thwart+Wal-Mart). *Wall Street Journal Online*. Archived from the original (http://pqasb.pqarchiver.com/wsj/access/1049173581.html?dids=1049173581:1049173581&FMT=ABS&FMTS=ABS:FT&date=Jun+6,+2006&author=Cecilie+Rohwedder&pub=Ms.+Fiala+Changes+Detergent;+Wall+Street+Journal&edition=Eastern+edition&startpage=A.1&type=91_1996&desc=Stores+of+Knowledge:+No.1+Retailer+in+Britain+Uses+'Clubcard'+to+Thwart+Wal-Mart) on 1 October 2007. . Retrieved 30 April 2007.(Abstract only is online)

[78] "Videotex takes Gateshead Teleshopping into the home". *The Incorporated Engineer, Journal of the IEEIE London*: 6. September 1984.

[79] "Tesco moves into software market" (http://news.bbc.co.uk/2/hi/business/5396488.stm). BBC News. 1 October 2006. .

[80] "Tesco Preliminary Results 2006" (http://www.tescocorporate.com/images/pressrelease_final_0.pdf). Tescocorporate.com. . Retrieved 16 August 2011.

[81] http://www.tescoplc.com/plc/ir/pres_results/analyst_packs/ap2010/prelim10/prelim10.pdf

[82] "TESCO Extends Partnership in China" (http://miranda.hemscott.com/servlet/HsPublic?context=ir.access&ir_option=RNS_NEWS&item=38141457174563&ir_client_id=622). London Stock Exchange's Regulatory News Service. 12 December 2006. .

[83] "Tescco Express rolls into China" (http://business.timesonline.co.uk/tol/business/industry_sectors/retailing/article3671705.ece). *The Times* (UK). 21 April 2008. .

[84] "Tesco Annual Review 2007" (http://www.tescocorporate.com/images/annual_review_and_sfs_2007_0.pdf) (PDF). Tesco Plc. 21 April 2008. .

[85] "Tesco Services" (http://en.itesco.cz/en_cz/services). Tesco Czech Republic. 21 April 2008. .
[86] "UK firms shut French stores as 'booze cruise' declines" (http://news.bbc.co.uk/1/hi/uk/10356727.stm). BBC News. 19 June 2010. . Retrieved 17 July 2010.
[87] "Tesco Services" (http://www.tesco.hu/tpp-main.php?cmssessid=T62c0c8b798583512629bddae51a154bad5f2f0682cea9b6a2185c8ab4d78185). Tesco Hungary. 21 April 2008. .
[88] "Tesco withdraws from Ireland". *The Guardian* (Guardian Newspapers). 27 March 1986.
[89] "Every little helps Tesco improve its bottom line – The Irish Times – Fri, Apr 01, 2011" (http://www.irishtimes.com/newspaper/finance/2011/0401/1224293541885.html). *The Irish Times*. 1 April 2011. . Retrieved 19 April 2011.
[90] "Food and Grocery Shopping, Tesco Ireland" (http://www.tesco.ie/online-shopping-ireland.html). Tesco Ireland. 21 April 2008. .
[91] "Tesco still commanding highest market share" (http://www.rte.ie/news/2005/1206/supermarkets-business.html). RTÉ Business. 6 December 2006. .
[92] "About Tesco Ireland" (http://www.tesco.ie/corporate_info/abouttesco.htm). Tesco Ireland. .
[93] "Tesco raises sourcing from India to £230 m" (http://www.financialexpress.com/news/Tesco-raises-sourcing-from-India-to--230-m/712253/). The Financial Express. 17 November 2010. . Retrieved 20 December 2010.
[94] "Tesco planning venture in India" (http://news.bbc.co.uk/2/hi/business/7555530.stm). BBC News. 12 August 2008. . Retrieved 29 December 2010.
[95] McCurry, Justin (1 June 2004). "Tesco pedals cautious path in Japan" (http://www.guardian.co.uk/business/2004/jun/01/japan.supermarkets). *The Guardian* (Guardian News and Media). . Retrieved 31 August 2011.
[96] "Tesco bids sayonara to 'unscalable' Japan business" (http://www.bbc.co.uk/news/business-14728099). *BBC News* (British Broadcasting Corporation). 31 August 2011. . Retrieved 31 August 2011.
[97] "Corporate information, Tesco Malaysia" (http://www.tesco.com.my/article.cfm?id=33). Tesco Stores (Malaysia)Sdn Bhd. 21 April 2008. .
[98] "Tesco: Interactive map" (http://www.tescoplc.com/plc/about_us/map/). Tescoplc.com. . Retrieved 16 April 2010.
[99] "Tesco will open four new stores, employing nearly 200 people" (http://translate.google.sk/translate?js=n&prev=_t&hl=sk&ie=UTF-8&layout=2&eotf=1&sl=sk&tl=en&u=http://ekonomika.sme.sk/c/5777349/tesco-otvori-styri-nove-predajne-zamestna-takmer-200-ludi.html). SME.sk. 22 February 2011. .
[100] "Tesco falls foul of Slovak government" (http://www.cee-foodindustry.com/news/ng.asp?id=67411-tesco-cee-slovakia). cee food industry. 2 May 2006. .
[101] "The new format Tesco Extra comes to Slovakia" (http://corporate.itesco.sk/tlacova-sprava.html?id=50). tesco slovakia. 30 March 2010. . In 2010 first Tesco Express stores were opened in Bratislava with current number of 10 shops. Tesco also operates store called 'My' in Bratislava which accept Clubcard and share some branding, however, most promotions do not apply for My.
[102] "E-Land in talk to sell stores to Tesco South Korea unit | Reuters" (http://uk.reuters.com/article/idUKSEO28008020080514). Uk.reuters.com. 14 May 2008. . Retrieved 16 April 2010.
[103] "Tesco to buy 36 S.Korean stores" (http://uk.reuters.com/article/mergersNews/idUKSEO28065320080514). Reuters. 14 May 2008. .
[104] "Tesco set to grow in South Korea" (http://news.bbc.co.uk/1/hi/business/7399957.stm). BBC News. 14 May 2008. . Retrieved 6 January 2010.
[105] "Key facts about Tesco Lotus" (http://www.tescolotus.net/company/keyfact.asp). Tesco Lotus. 21 April 2008. .
[106] "Tesco to enter United States" (http://www.tescocorporate.com/page.aspx?pointerid=14163CB2412F41B1BD7765AC8DBE49EB). . Retrieved 13 March 2006.
[107] Wearden, Graeme (20 April 2010). "Tesco rings up record profits" (http://www.guardian.co.uk/business/2010/apr/20/tesco-rings-up-record-profits-again). London: Guardian News and Media Limited. . Retrieved 20 April 2010.
[108] Wood, Zoe (2011-10-05). "Tesco's UK sales slide as consumers cut non-essential spending" (http://www.guardian.co.uk/business/2011/oct/05/tesco-sales-slide-consumers-cut-back). *The Guardian* (London). . Retrieved October 05, 2011. "Tesco has reported its weakest six-monthly UK sales figures for 20 years as higher food and fuel costs contributed to stark decline in spending on non-essentials such as gadgets, CDs and games in its stores."
[109] "Tesco UK Grocery Market Share Up At 30.5% 12 Weeks To Dec 27 – TNS" (http://www.advfn.com/lse/ShareNews.asp?sharenews=TSCO&article=41054419&headline=tesco-uk-grocery-market-share-up-at-30-5-12-weeks-to-dec-27). ADVFN. .
[110] Evans, Rob (30 April 2008). "Writers criticise Tesco for 'chilling' Thai libel actions" (http://www.guardian.co.uk/business/2008/apr/30/tesco.supermarkets). *The Guardian* (London). .
[111] "UK grocers face competition probe" (http://news.bbc.co.uk/1/hi/business/6287923.stm). BBC. 22 January 2007. . Retrieved 21 February 2007.
[112] "Tesco 'breaching planning laws'" (http://news.bbc.co.uk/1/hi/uk/5261844.stm). BBC. 18 August 2006. . Retrieved 21 February 2007.
[113] "Tesco Till 'Slowest'" (http://www.mirror.co.uk/news/tm_headline=tesco-till--slowest--&method=full&objectid=18289622&siteid=94762-name_page.html). *The Mirror*. 20 December 2006. . Retrieved 22 December 2006.
[114] "Tesco.com Pricecheck" (http://www.tesco.com/todayattesco/pricecheck.shtml). *Tesco* (Tesco). 11 July 2007. . Retrieved 13 July 2007.
[115] Thompson, James (6 October 2009). "Diageo ends link with US TV host but Tesco stays" (http://www.independent.co.uk/news/business/news/diageo-ends-link-with-us-tv-host-but-tesco-stays-1798270.html). *The Independent* (London). . Retrieved 6 October 2009.

[116] Andrew Foxwell and Lauren Mills (13 May 2007). "'Green' Tesco's Swiss tax dodge" (http://www.thisismoney.co.uk/consumer/caring/article.html?in_article_id=420259&in_page_id=511). Mail on Sunday. .

[117] David Leigh (14 June 2008). "Government outlaws tax avoidance schemes" (http://www.guardian.co.uk/business/2008/jun/14/taxavoidance.tesco). *The Guardian* (UK). .

[118] Richard Murphy (1 June 2008). "Tescos: the Zug deal is tax avoidance" (http://www.taxresearch.org.uk/Blog/2008/06/01/tescos-the-zug-deal-is-tax-avoidance/). .

[119] Ian Griffiths and Felicity Lawrence (27 February 2008). "Tesco's £1bn tax avoiding plan – move to the Cayman Islands (legal action/article removed from Guardian web site)" (http://www.tiscali.co.uk/money/guardian/news/2008/02/27/tescos-1631bn-tax-avoiding-plan-move-to-t.html). *The Guardian* (UK). Archived from the original (http://www.guardian.co.uk/business/2008/feb/27/tesco.supermarkets) on 27 February 2008. .

[120] Patrick Wintour (28 February 2008). "(legal action/article removed from Guardian web site)" (http://www.guardian.co.uk/business/2008/feb/28/tesco.supermarkets?gusrc=rss&feed=networkfront). *The Guardian* (UK). .

[121] "Tesco and tax: a complex web of companies, trusts and partnerships" (http://www.guardian.co.uk/business/2008/may/03/tesco.medialaw). *The Guardian* (UK). 3 May 2008. .

[122] "Corrections and clarifications – Tesco – an apology" (http://www.guardian.co.uk/theguardian/2008/sep/16/corrections). *The Guardian* (UK). 16 September 2008. .

[123] Daniel Farey-Jones (16 September 2008). "Guardian apologises and pays damages to settle Tesco tax case" (http://www.brandrepublic.com/News/846169/Guardian-apologises-pays-damages-settle-Tesco-tax-case/). Brand Republic. .

[124] "Experts warn of Tesco expansion" (http://www.thisismoney.co.uk/news/article.html?in_article_id=408450&in_page_id=2). *Evening Standard*. UK. 21 April 2006. .

[125] "Michael Wilkes " 2009 " January" (http://michaelwilkes.mycouncillor.org.uk/2009/01/). Michaelwilkes.mycouncillor.org.uk. . Retrieved 16 April 2010.

[126] "Mixed reaction to Tesco expansion at Cannon Park, Coventry" (http://www.coventrytelegraph.net/news/coventry-news/2008/07/18/mixed-reaction-to-tesco-expansion-at-cannon-park-coventry-92746-21364275/). Coventry Telegraph. 18 July 2008. .

[127] "No Mill Road Tesco" (http://www.nomillroadtesco.org/). the No Mill Road Tesco Campaign. . Retrieved 16 April 2010.

[128] "Tesco appeals dismissed over Mill Road store" (http://www.cambridge-news.co.uk/cn_news_home/DisplayArticle.asp?ID=365818). Cambridge Evening News. 12 November 2008. .

[129] "Appeal Decisions: Inquiry held on 30 September – 3 October 2008" (http://www.nomillroadtesco.org/wp-content/uploads/2008/11/appeal1result.pdf) (PDF). . Retrieved 16 April 2010.

[130] "Denied: Tesco's Mill Road plan fails test" (http://www.cambridge-news.co.uk/cn_news_home/DisplayArticle.asp?ID=336029). Cambridge Evening News. 1 August 2008. .

[131] "New Tesco opens despite protesters" (http://www.cambridge-news.co.uk/cn_news_cambridge/displayarticle.asp?id=444630). Cambridge Evening News. 27 August 2009. .

[132] Kirby, Terry (27 March 2007). "Teetotal Bournville takes pledge against Tesco – This Britain, UK" (http://www.independent.co.uk/news/uk/this-britain/teetotal-bournville-takes-pledge-against-tesco-442048.html). *The Independent* (UK). . Retrieved 16 April 2010.

[133] "Tesco may appeal" (http://www.stalbansreview.co.uk/news/2371862.tesco_may_appeal/). St Albans Review. 30 June 2008. .

[134] Moore, Alex; Leader, Rhondda (8 January 2009). "'Tesco will make us a ghost town'" (http://www.walesonline.co.uk/news/south-wales-news/rhondda/2009/01/08/tesco-will-make-us-a-ghost-town-91466-22618667/). WalesOnline. .

[135] "Tesco Store Locator" (http://www.tesco.com/storeLocator/). Tesco Online. 2 September 2009. .

[136] "No Tesco on Hope Street Poster" (http://hopestreet.co.uk/2009/08/poster-no-tesco-on-hope-street/). Hope Street Online. 2 September 2009. .

[137] "There is just something plain WRONG about a Tesco on Hope Street" (https://www.facebook.com/group.php?gid=259254990413&ref=ts). *Facebook*. 2 September 2009. .

[138] "Police hurt in violent anti-Tesco clashes in Bristol" (http://www.bbc.co.uk/news/uk-england-bristol-13167041). BBC News. 22 April 2011. . Retrieved 22 April 2011.

[139] Jane Onyanga-Omara (22 April 2011). "Chaos in Bristol's 'cultural quarter'" (http://www.bbc.co.uk/news/uk-england-bristol-13169684). BBC News. . Retrieved 22 April 2011.

[140] "Council welcomes proposal for Tesco supermarket at Kennedy Way, Immingham" (http://www.nelincs.gov.uk/news/tescoimmingham.htm). North East Lincolnshire Council. 29 January 2009. .

[141] "New Tesco plan is unveiled" (http://www.thisisgrimsby.co.uk/news/Shopping-development-big-boost-Immingham/article-653146-detail/article.html). This is Grimsby. 29 January 2009. .

[142] "FT.Com, Telecoms – New chief rings the changes at TalkTalk" (http://www.ft.com/cms/s/0/c34b13a8-ff1d-11df-956b-00144feab49a.html). *Financial Times*. 3 December 2010. . Retrieved 19 April 2011.

[143] "Retail – Halfords slips a gear as cycle sales back-pedal" (http://www.ft.com/cms/s/5d64773e-d22f-11df-8fbe-00144feabdc0,Authorised=false.html?_i_location=http://www.ft.com/cms/s/0/5d64773e-d22f-11df-8fbe-00144feabdc0.html&_i_referer=http://search.ft.com/search?queryText=David+Wild&x=0&y=0&aje=false&dse=&dsz=). *Financial Times*. . Retrieved 19 April 2011.

[144] "Travel & Leisure – Reynolds joins list of Tesco leavers" (http://www.ft.com/cms/s/6b48e64c-7dae-11dc-9f47-0000779fd2ac,Authorised=false.html?_i_location=http://www.ft.com/cms/s/0/

6b48e64c-7dae-11dc-9f47-0000779fd2ac.html&_i_referer=http://search.ft.com/search?queryText=Julia+Reynolds&x=0&y=0&aje=false&dse=&dsz=). *Financial Times*. 19 October 2007. . Retrieved 19 April 2011.

[145] "UK – Browett to cultivate consistency at DSG" (http://www.ft.com/cms/s/6cf8d724-a467-11dc-a28d-0000779fd2ac,Authorised=false.html?_i_location=http://www.ft.com/cms/s/0/6cf8d724-a467-11dc-a28d-0000779fd2ac.html&_i_referer=http://search.ft.com/search?queryText=John+Browett&x=0&y=0&aje=false&dse=&dsz=). *Financial Times*. 7 December 2007. . Retrieved 19 April 2011.

[146] "People – New Look's new chairman" (http://www.ft.com/cms/s/4f883926-e076-11de-8494-00144feab49a,Authorised=false.html?_i_location=http://www.ft.com/cms/s/0/4f883926-e076-11de-8494-00144feab49a.html&_i_referer=http://search.ft.com/search?queryText=John+Gildersleeve&x=0&y=0&aje=false&dse=&dsz=). *Financial Times*. 4 December 2009. . Retrieved 19 April 2011.

[147] "People – French shopping" (http://www.ft.com/cms/s/3b35df20-0176-11df-8c54-00144feabdc0,Authorised=false.html?_i_location=http://www.ft.com/cms/s/0/3b35df20-0176-11df-8c54-00144feabdc0.html&_i_referer=http://search.ft.com/search?queryText=James+McCann&x=0&y=0&aje=false&dse=&dsz=). *Financial Times*. 15 January 2010. . Retrieved 19 April 2011.

[148] "companies, Travel & leisure – McMeikan mulls Sainsbury's performance" (http://www.ft.com/cms/s/ad4fba2a-6ea1-11da-9544-0000779e2340,Authorised=false.html?_i_location=http://www.ft.com/cms/s/0/ad4fba2a-6ea1-11da-9544-0000779e2340.html&_i_referer=http://search.ft.com/search?queryText=Ken+Mcmeikan&x=0&y=0&aje=false&dse=&dsz=). *Financial Times*. 17 December 2005. . Retrieved 19 April 2011.

[149] Thompson, James (19 January 2011). "Domino's Pizza lines up Lance Batchelor as next chief executive" (http://www.independent.co.uk/news/business/news/dominos-pizza-lines-up-lance-batchelor-as-next-chief-executive-2188629.html). *The Independent* (London). .

External links

Official

- Tesco UK (http://www.tesco.com)
- Tesco PLC (http://www.tescoplc.com/)

Other

- Tesco companies (http://opencorporates.com/corporate_groupings/tesco) grouped at OpenCorporates
- Tescopoly.org (http://www.tescopoly.org/), Coalition of campaign groups criticising Tesco.

Pixel

In digital imaging, a **pixel**, or **pel**,[1] (**picture element**[2]) is a single point in a raster image, or the smallest addressable screen element in a display device; it is the smallest unit of picture that can be represented or controlled.

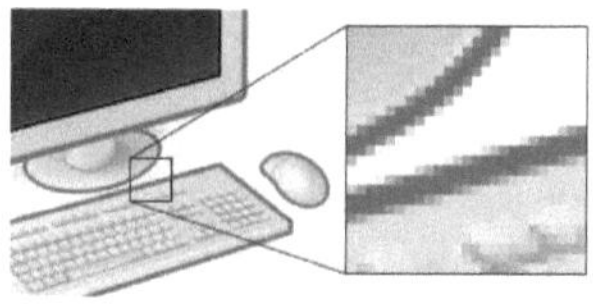

This example shows an image with a portion greatly enlarged, in which the individual pixels are rendered as little squares and can easily be seen.

Each pixel has its own address. The address of a pixel corresponds to its coordinates. Pixels are normally arranged in a two-dimensional grid, and are often represented using dots or squares. Each pixel is a sample of an original image; more samples typically provide more accurate representations of the original. The intensity of each pixel is variable. In color image systems, a color is typically represented by three or four component intensities such as red, green, and blue, or cyan, magenta, yellow, and black.

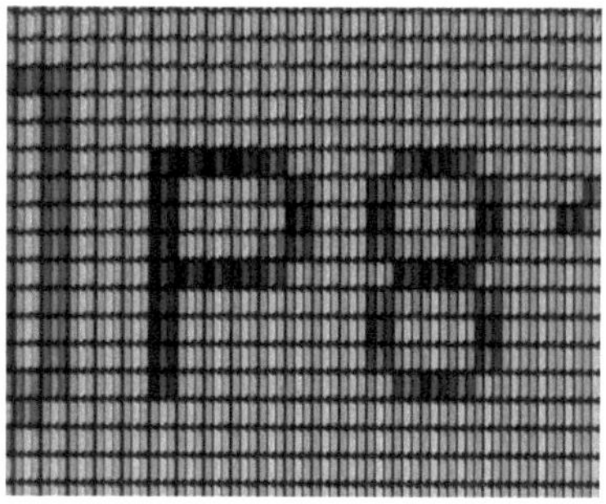

A photograph of sub-pixel display elements on a laptop's LCD screen

In some contexts (such as descriptions of camera sensors), the term *pixel* is used to refer to a single scalar element of a multi-component representation (more precisely called a *photosite* in the camera sensor context, although the neologism *sensel* is sometimes used to describe the elements of a digital camera's sensor),[3] while in others the term may refer to the entire set of such component intensities for a spatial position. In color systems that use chroma subsampling, the multi-component concept of a pixel can become difficult to apply, since the intensity measures for the different color components correspond to different spatial areas in a such a representation.

The word *pixel* is based on a contraction of *pix* ("pictures") and *el* (for "element"); similar formations with *el* for "element" include the words voxel[4] and texel.[4]

Etymology

The word "pixel" was first published in 1965 by Frederic C. Billingsley of JPL, to describe the picture elements of video images from space probes to the Moon and Mars. However, Billingsley did not coin the term himself. Instead, he got the word "pixel" from Keith E. McFarland, at the Link Division of General Precision in Palo Alto, who did not know where the word originated. McFarland said simply it was "in use at the time" (circa 1963).[5]

The word is a combination of *picture* and *element*, via *pix*. The word *pix* appeared in *Variety* magazine headlines in 1932, as an abbreviation for the word *pictures*, in reference to movies.[6] By 1938, "pix" was being used in reference to still pictures by photojournalists.[5]

The concept of a "picture element" dates to the earliest days of television, for example as "*Bildpunkt*" (the German word for *pixel*, literally 'picture point') in the 1888 German patent of Paul Nipkow. According to various etymologies, the earliest publication of the term *picture element* itself was in *Wireless World* magazine in 1927,[7] though it had been used earlier in various U.S. patents filed as early as 1911.[8]

Some authors explain *pixel* as *picture cell,* as early as 1972.[9] In video processing, *pel* is often used instead of *pixel*.[10] For example, IBM used it in their Technical Reference for the original PC.

Words with similar etymologies

- Texel (texture element) and luxel (lux element) are words used to describe a pixel when it is used in specific context (texturing and light mapping respectively)
- A voxel is a *volume element*, the 3D analogue of a 2D pixel.
- Surfels (surface elements) have the same naming pattern as pixels, but share more similarities with shrunken triangles than expanded pixels.
- A resel is a *resolution element*, used to describe the discrete number of resolvable elements of a continuous signal (e.g. in optical imaging).

Technical

A pixel is generally thought of as the smallest single component of a digital image. However, the definition is highly context-sensitive. For example, there can be "printed pixels" in a page, or pixels carried by electronic signals, or represented by digital values, or pixels on a display device, or pixels in a digital camera (photosensor elements). This list is not exhaustive, and depending on context, there are several terms that are synonymous in particular contexts, such as pel, sample, byte, bit, dot, spot, etc. The term "pixels" can be used in the abstract, or as a unit of measure, in particular when using pixels as a measure of resolution, such as: 2400 pixels per inch, 640 pixels per line, or spaced 10 pixels apart.

A pixel does not need to be rendered as a small square. This image shows alternative ways of reconstructing an image from a set of pixel values, using dots, lines, or smooth filtering.

The measures dots per inch (dpi) and pixels per inch (ppi) are sometimes used interchangeably, but have distinct meanings, especially for printer devices, where dpi is a measure of the printer's density of dot (e.g. ink droplet) placement.[11] For example, a high-quality photographic image may be printed with 600 ppi on a 1200 dpi inkjet printer.[12] Even higher dpi numbers, such as the 4800 dpi quoted by printer manufacturers since 2002, do not mean much in terms of achievable resolution.[13]

The more pixels used to represent an image, the closer the result can resemble the original. The number of pixels in an image is sometimes called the resolution, though resolution has a more specific definition. Pixel counts can be expressed as a single number, as in a "three-megapixel" digital camera, which has a nominal three million pixels, or as a pair of numbers, as in a "640 by 480 display", which has 640 pixels from side to side and 480 from top to bottom (as in a VGA display), and therefore has a total number of $640 \times 480 = 307{,}200$ pixels or 0.3 megapixels.

The pixels, or color samples, that form a digitized image (such as a JPEG file used on a web page) may or may not be in one-to-one correspondence with screen pixels, depending on how a computer displays an image. In computing, an image composed of pixels is known as a *bitmapped image* or a *raster image*. The word *raster* originates from television scanning patterns, and has been widely used to describe similar halftone printing and storage techniques.

Sampling patterns

For convenience, pixels are normally arranged in a regular two-dimensional grid. By using this arrangement, many common operations can be implemented by uniformly applying the same operation to each pixel independently. Other arrangements of pixels are possible, with some sampling patterns even changing the shape (or kernel) of each pixel across the image. For this reason, care must be taken when acquiring an image on one device and displaying it on another, or when converting image data from one pixel format to another.

For example:

- LCD screens typically use a staggered grid, where the red, green, and blue components are sampled at slightly different locations. Subpixel rendering is a technology which takes advantage of these differences to improve the rendering of text on LCD screens.

Text rendered using ClearType

- The vast majority of color digital cameras use a Bayer filter, resulting in a regular grid of pixels where the *color* of each pixel depends on its position on the grid.
- A clipmap uses a hierarchical sampling pattern, where the size of the support of each pixel depends on its location within the hierarchy.
- Warped grids are used when the underlying geometry is non-planar, such as images of the earth from space.[14]
- The use of non-uniform grids is an active research area, attempting to bypass the traditional Nyquist limit.[15]
- Pixels on computer monitors are normally "square" (this is, having equal horizontal and vertical sampling pitch); pixels in other systems are often "rectangular" (that is, having unequal horizontal and vertical sampling pitch – oblong in shape), as are digital video formats with diverse aspect ratios, such as the anamorphic widescreen formats of the Rec. 601 digital video standard.

Resolution of computer monitors

Computers can use pixels to display an image, often an abstract image that represents a GUI. The resolution of this image is called the display resolution and is determined by the video card of the computer. LCD monitors also use pixels to display an image, and have a native resolution. Each pixel is made up of triads, with the number of these triads determining the native resolution. On some CRT monitors, the beam sweep rate may be fixed, resulting in a fixed native resolution. Most CRT monitors do not have a fixed beam sweep rate, meaning they do not have a native resolution at all - instead they have a set of resolutions that are equally well supported. To produce the sharpest images possible on an LCD, the user must ensure the display resolution of the computer matches the native resolution of the monitor.

Bits per pixel

The number of distinct colors that can be represented by a pixel depends on the number of bits per pixel (bpp). A 1 bpp image uses 1-bit for each pixel, so each pixel can be either on or off. Each additional bit doubles the number of colors available, so a 2 bpp image can have 4 colors, and a 3 bpp image can have 8 colors:

- 1 bpp, $2^1 = 2$ colors (monochrome)
- 2 bpp, $2^2 = 4$ colors
- 3 bpp, $2^3 = 8$ colors

  ...

- 8 bpp, $2^8 = 256$ colors
- 16 bpp, $2^{16} = 65{,}536$ colors ("Highcolor")
- 24 bpp, $2^{24} \approx 16.8$ million colors ("Truecolor")

For color depths of 15 or more bits per pixel, the depth is normally the sum of the bits allocated to each of the red, green, and blue components. Highcolor, usually meaning 16 bpp, normally has five bits for red and blue, and six bits for green, as the human eye is more sensitive to errors in green than in the other two primary colors. For applications involving transparency, the 16 bits may be divided into five bits each of red, green, and blue, with one bit left for transparency. A 24-bit depth allows 8 bits per component. On some systems, 32-bit depth is available: this means that each 24-bit pixel has an extra 8 bits to describe its opacity (for purposes of combining with another image).

Subpixels

Many display and image-acquisition systems are, for various reasons, not capable of displaying or sensing the different color channels at the same site. Therefore, the pixel grid is divided into single-color regions that contribute to the displayed or sensed color when viewed at a distance. In some displays, such as LCD, LED, and plasma displays, these single-color regions are separately addressable elements, which have come to be known as *subpixels*. For example, LCDs typically divide each pixel horizontally into three subpixels. When the square pixel is divided into three subpixels, each subpixel is necessarily rectangular. In the display industry terminology, subpixels are often referred to as *pixels*, as they are the basic addressable elements in a viewpoint of hardware, and they call *pixel circuits* rather than *subpixel circuits*.

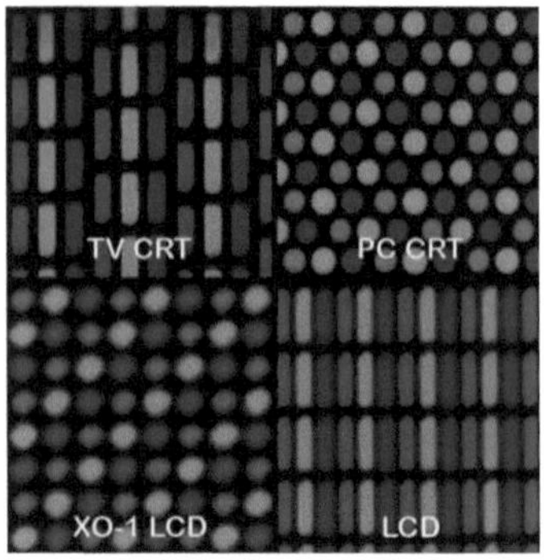

Geometry of color elements of various CRT and LCD displays; phosphor dots in a color CRT display (top row) bear no relation to pixels or subpixels.

Most digital camera image sensors use single-color sensor regions, for example using the Bayer filter pattern, and in the camera industry these are known as *pixels* just like in the display industry, not *subpixels*.

For systems with subpixels, two different approaches can be taken:

- The subpixels can be ignored, with full-color pixels being treated as the smallest addressable imaging element; or
- The subpixels can be included in rendering calculations, which requires more analysis and processing time, but can produce apparently superior images in some cases.

This latter approach, referred to as subpixel rendering, uses knowledge of pixel geometry to manipulate the three colored subpixels separately, producing a slight increase in the apparent resolution of color displays. While CRT displays use red-green-blue-masked phosphor areas, dictated by a mesh grid called the shadow mask, it would require a difficult calibration step to be aligned with the displayed pixel raster, and so CRTs do not currently use subpixel rendering.

The concept of subpixels is related to samples.

Megapixel

A megapixel (MP or Mpx) is one million pixels, and is a term used not only for the number of pixels in an image, but to express the number of image sensor elements of digital cameras or the number of display elements of digital displays. For example, a camera with an array of 2048 × 1536 sensor elements is commonly said to have "3.1 megapixels" (2048 × 1536 = 3,145,728).

Digital cameras use photosensitive electronics, either charge-coupled device (CCD) or complementary metal–oxide–semiconductor (CMOS) image sensors, consisting of a large number of single sensor elements, each of which records a measured intensity level. In most digital cameras, the sensor array is covered with a patterned color filter mosaic having red, green, and blue regions in the Bayer filter arrangement, so that each sensor element can record the intensity of a single primary color of light. The camera interpolates the color information of neighboring sensor elements, through a process called demosaicing, to create the final image. These sensor elements are often called "pixels", even though they only record 1 channel (only red, or green, or blue) of the final color image. Thus, two of the three color channels for each sensor must be interpolated and a so-called *N-megapixel* camera that produces an N-megapixel image provides only one-third of the information that an image of the same size could get from a scanner. Thus, certain color contrasts may look fuzzier than others, depending on the allocation of the primary colors (green has twice as many elements as red or blue in the Bayer arrangement).

Display resolution

The display resolution of a display device (such as a digital television) is the number of distinct pixels in each dimension that can be displayed. It can be an ambiguous term especially as the displayed resolution is controlled by different factors in cathode ray tube (CRT) and flat panel or projection displays using fixed-element arrays.

See also

- Computer display standard
- Dexel
- Gigapixel image
- Image resolution
- Intrapixel and Interpixel processing
- LCD crosstalk
- Pixel advertising
- Pixel art
- Pixel art scaling algorithms
- Pixel aspect ratio
- Point (typography)
- Rasterisation
- Vector graphics
- Voxel

References

[1] J.D Foley and A. Van Dam, Fundamentals of Interactive Computer Graphics, 1982

[2] Rudolf F. Graf (1999). *Modern Dictionary of Electronics* (http://books.google.com/?id=o2I1JWPpdusC&pg=PA569). Oxford: Newnes. p. 569. ISBN 0-7506-43315. .

[3] Michael Goesele (2004). *New Acquisition Techniques for Real Objects and Light Sources in Computer Graphics* (http://books.google.com/?id=ZTJJ8QzNv1wC&pg=PA33&dq=sensel+sensor-element). Books on Demand. ISBN 3833414898. .

[4] Foley, James D.; Andries van Dam, John F. Hughes, Steven K. Feiner (1990). "Spatial-partitioning representations; Surface detail". *Computer Graphics: Principles and Practice*. The Systems Programming Series. Addison-Wesley. ISBN 0-201-12110-7. "These cells are often called *voxels* (volume elements), in analogy to pixels."

[5] Lyon, Richard F. (2006). A brief history of 'pixel' (http://www.foveon.com/files/ABriefHistoryofPixel2.pdf). IS&T/SPIE Symposium on Electronic Imaging.

[6] "Online Etymology Dictionary" (http://www.etymonline.com/index.php?search=pixel). .

[7] "On language; Modem, I'm Odem" (http://query.nytimes.com/gst/fullpage.html?res=990CE3DB1E31F931A35757C0A963958260), *The New York Times*, April 2, 1995. Accessed April 7, 2008.

[8] Sinding-Larsen, Alf Transmission of Pictures of Moving Objects (http://www.google.com/patents?vid=USPAT1175313&id=Df9EAAAAEBAJ&dq="picture+element"&as_drrb_ap=b), US Patent 1,175,313, issued March 14, 1916.

[9] Robert L. Lillestrand (1972). "Techniques for Change Detection". *IEEE Trans. Computers* **C-21** (7).

[10] Lewis, Peter H. (February 12, 1989). The Executive Computer; Compaq Sharpens Its Video Option (http://www.nytimes.com/1989/02/12/business/the-executive-computer-compaq-sharpens-its-video-option.html). *The New York Times*.

[11] Derek Doeffinger (2005). *The Magic of Digital Printing* (http://books.google.com/?id=s2hIx1amJUcC&dq=printer+dots-per-inch+pixels-per-inch). Lark Books. p. 24. ISBN 1579906893. .

[12] "Experiments with Pixels Per Inch (PPI) on Printed Image Sharpness" (http://www.clarkvision.com/imagedetail/printer-ppi/). *ClarkVision.com*. July 3, 2005. .

[13] Harald Johnson (2002). *Mastering Digital Printing* (http://books.google.com/?id=wto19gxFyfQC&dq=inkjet+printer+4800+dpi+addressability). Thomson Course Technology. p. 40. ISBN 1929685653. .

[14] "Image registration of blurred satellite images" (http://staff.utia.cas.cz/zitova/registration.htm). . Retrieved 2008-05-09.

[15] "ScienceDirect - Pattern Recognition: Image representation by a new optimal non-uniform morphological sampling:" (http://www.sciencedirect.com/science/article/B6V14-3YXB1T5-8/2/15b559f2aefbc7bdc54c55df148c2374). . Retrieved 2008-05-09.

External links

- A Pixel Is Not A Little Square (http://alvyray.com/Memos/CG/Microsoft/6_pixel.pdf): Microsoft Memo by computer graphics pioneer Alvy Ray Smith.
- Pixels and Me (http://video.google.com/videoplay?docid=-5655850487750051532): Video of a history talk at the Computer History Museum.
- Square and non-Square Pixels (http://lurkertech.com/lg/pixelaspect/): Technical info on pixel aspect ratios of modern video standards (480i,576i,1080i,720p), plus software implications.
- 120 Megapixel is here now (http://www.gigapixel360.com): A lot of information about MegaPixel and Gigapixel.

Digital camera

A **digital camera** (or **digicam**) is a camera that takes video or still photographs, or both, digitally by recording images via an electronic image sensor. It is the main device used in the field of digital photography. Most 21st century cameras are digital.[1]

Front and back of Canon PowerShot A95

Digital cameras can do things film cameras cannot: displaying images on a screen immediately after they are recorded, storing thousands of images on a single small memory device, and deleting images to free storage space. The majority, including most compact cameras, can record moving video with sound as well as still photographs. Some can crop and stitch pictures and perform other elementary image editing. Some have a GPS receiver built in, and can produce Geotagged photographs.

The optical system works the same as in film cameras, typically using a lens with a variable diaphragm to focus light onto an image pickup device. The diaphragm and shutter admit the correct amount of light to the imager, just as with film but the image pickup device is electronic rather than chemical. Most digicams, apart from camera phones and a few specialized types, have a standard tripod screw.

Digital cameras are incorporated into many devices ranging from PDAs and mobile phones (called camera phones) to vehicles. The Hubble Space Telescope and other astronomical devices are essentially specialized digital cameras.

Types of digital cameras

Digital cameras are made in a wide range of sizes, prices and capabilities. The majority are camera phones, operated as a mobile application through the cellphone menu. Professional photographers and many amateurs use larger, more expensive digital single-lens reflex cameras (DSLR) for their greater versatility. Between these extremes lie digital compact cameras and bridge digital cameras that "bridge" the gap between amateur and professional cameras. Specialized cameras including multispectral imaging equipment and astrographs continue to serve the scientific, military, medical and other special purposes for which digital photography was invented.

Compact digital cameras

Compact cameras are designed to be tiny and portable and are particularly suitable for casual and "snapshot" uses. Hence, they are also called point-and-shoot cameras. The smallest, generally less than 20 mm thick, are described as *subcompacts* or "ultra-compacts" and some are nearly credit card size.[2]

Subcompact with lens assembly retracted

Most, apart from ruggedized or water-resistant models, incorporate a retractable lens assembly allowing a thin camera to have a moderately long focal length and thus fully exploit an image sensor larger than that on a camera phone, and a mechanized lens cap to cover the lens when retracted. The retracted and capped lens is protected from keys, coins and other hard objects, thus making it a thin, pocketable package. Subcompacts commonly have one lug and a short wrist strap which aids extraction from a pocket, while thicker compacts may have two lugs for attaching a neck strap.

Compact cameras are usually designed to be easy to use, sacrificing advanced features and picture quality for compactness and simplicity; images can usually only be stored using lossy compression (JPEG). Most have a built-in flash usually of low power, sufficient for nearby subjects. Live preview is almost always used to frame the photo. Most have limited motion picture capability. Compacts often have macro capability and zoom lenses but the zoom range is usually less than for bridge and DSLR cameras. Generally a contrast-detect autofocus system, using the image data from the live preview feed of the main imager, focuses the lens.

Typically, these cameras incorporate a nearly silent leaf shutter into their lenses.

For lower cost and smaller size, these cameras typically use image sensors with a diagonal of approximately 6 mm, corresponding to a crop factor around 6. This gives them weaker low-light performance, greater depth of field, generally closer focusing ability, and smaller components than cameras using larger sensors.

Starting in 2011, some compact digital cameras can take 3D still photos. These 3D compact stereo cameras can capture 3D panoramic photos for play back on a 3D TV.[3] Some of these are rugged and waterproof, and some have GPS, compass, barometer and altimeter. [4]

Bridge cameras

Bridge are higher-end digital cameras that physically and ergonomically resemble DSLRs and share with them some advanced features, but share with compacts the use of a fixed lens and a small sensor. Like compacts, most use live preview to frame the image. Their autofocus uses the same contrast-detect mechanism, but many bridge cameras have a manual focus mode, in some cases using a separate focus ring, for greater control. They originally "bridged" the gap between affordable point-and-shoot cameras and the then unaffordable earlier digital SLRs.

Sony DSC-H2

Due to the combination of big physical size but a small sensor, many of these cameras have very highly specified lenses with large zoom range and fast aperture, partially compensating for the inability to change lenses. On some, the lens qualifies as superzoom. To compensate for the lesser sensitivity of their small sensors, these cameras almost always include an image stabilization system to enable longer handheld exposures.

These cameras are sometimes marketed as and confused with digital SLR cameras since the appearance is similar. Bridge cameras lack the reflex viewing system of DSLRs, are usually fitted with fixed (non-interchangeable) lenses (although some have a lens thread to attach accessory wide-angle or telephoto converters), and can usually take movies with sound. The scene is composed by viewing either the liquid crystal display or the electronic viewfinder (EVF). Most have a longer shutter lag than a true dSLR, but they are capable of good image quality (with sufficient light) while being more compact and lighter than DSLRs. High-end models of this type have comparable resolutions to low and mid-range DSLRs. Many of these cameras can store images in a Raw image format, or processed and JPEG compressed, or both. The majority have a built-in flash similar to those found in DSLRs.

In bright sun, the quality difference between a good compact camera and a digital SLR is minimal but bridgecams are more portable, cost less and have a similar zoom ability to dSLR. Thus a Bridge camera may better suit outdoor daytime activities, except when seeking professional-quality photos.[5]

In low light conditions and/or at ISO equivalents above 800, most bridge cameras (or megazooms) lack in image quality when compared to even entry level DSLRs. However, they do have one major advantage: their much larger depth of field due to the small sensor as compared to a DSLR, allowing larger apertures with shorter exposure times.

A 3D Photo Mode was introduced in 2011, whereby the camera automatically takes a second image from a slightly different perspective and provides a standard .MPO file for stereo display. [6]

Mirrorless interchangeable-lens camera

In late 2008, a new type of camera emerged, combining the larger sensors and interchangeable lenses of DSLRs with the live-preview viewing system of compact cameras, either through an electronic viewfinder or on the rear LCD. These are simpler and more compact than DSLRs due to the removal of the mirror box, and typically emulate the handling and ergonomics of either DSLRs or compacts. The system is used by Micro Four Thirds, borrowing components from the Four Thirds DSLR system.

Digital single lens reflex cameras

Digital single-lens reflex cameras (DSLRs) are digital cameras based on film single-lens reflex cameras (SLRs). They take their name from their unique viewing system, in which a mirror reflects light from the lens through a separate optical viewfinder. At the moment of exposure the mirror flips out of the way, making a distinctive "clack" sound and allowing light to fall on the imager.

Cutaway of an Olympus E-30 DSLR

Since no light reaches the imager during framing, autofocus is accomplished using specialized sensors in the mirror box itself. Most 21st century DSLRs also have a "live view" mode that emulates the live preview system of compact cameras, when selected.

These cameras have much larger sensors than the other types, typically 18 mm to 36 mm on the diagonal (crop factor 2, 1.6, or 1). This gives them superior low-light performance, less depth of field at a given aperture, and a larger size.

They make use of interchangeable lenses; each major DSLR manufacturer also sells a line of lenses specifically intended to be used on their cameras. This allows the user to select a lens designed for the application at hand: wide-angle, telephoto, low-light, etc. So each lens does not require its own shutter, DSLRs use a focal-plane shutter in front of the imager, behind the mirror.

Digital rangefinders

A rangefinder is a user-operated optical mechanism to measure subject distance once widely used on film cameras. Most digital cameras measure subject distance automatically using electro-optical techniques, but it is not customary to say that they have a rangefinder.

Line-scan camera systems

A line-scan camera is a camera device containing a line-scan image sensor chip, and a focusing mechanism. These cameras are almost solely used in industrial settings to capture an image of a constant stream of moving material. Unlike video cameras, line-scan cameras use a single row of pixel sensors, instead of a matrix of them. Data coming from the line-scan camera has a frequency, where the camera scans a line, waits, and repeats. The data coming from the line-scan camera is commonly processed by a computer, to collect the one-dimensional line data and to create a two-dimensional image. The collected two-dimensional image data is then processed by image-processing methods for industrial purposes.

Further information: Rotating line camera

Integration

Many devices include digital cameras built into or integrated into them. For example, mobile phones often include digital cameras; those that do are known as camera phones. Other small electronic devices (especially those used for communication) such as PDAs, laptops and BlackBerry devices often contain an integral digital camera, and most 21st century camcorders can also make still pictures.

Due to the limited storage capacity and general emphasis on convenience rather than image quality, almost all these integrated or converged devices store images in the lossy but compact JPEG file format.

Mobile phones incorporating digital cameras were introduced in Japan in 2001 by J-Phone. In 2003 camera phones outsold stand-alone digital cameras, and in 2006 they outsold all film-based cameras and digital cameras combined. These camera phones reached a billion devices sold in only five years, and by 2007 more than half of the installed base of all mobile phones were camera phones. Sales of separate cameras peaked in 2008. [7]

Integrated cameras tend to be at the very lowest end of the scale of digital cameras in technical specifications, such as resolution, optical quality, and ability to use accessories. With rapid development, however, the gap between mainstream compact digital cameras and camera phones is closing, and high-end camera phones are competitive with low-end stand-alone digital cameras of the same generation.

History

Steven Sasson as an engineer at Eastman Kodak invented and built the first digital camera using a charge-coupled device image sensor in 1975.[8] [9]

Image sensors

Further information: Image sensor

Image resolution

The resolution of a digital camera is often limited by the image sensor (typically a CCD or CMOS sensor chip) that turns light into discrete signals, replacing the job of film in traditional photography. The sensor is made up of millions of "buckets" that essentially count the number of photons that strike the sensor. This means that the brighter the image at a given point on the sensor, the larger the value that is read for that pixel. Depending on the physical structure of the sensor, a color filter array may be used which requires a demosaicing/interpolation algorithm. The number of resulting pixels in the image determines its "pixel count". For example, a 640x480 image would have 307,200 pixels, or approximately 307 kilopixels; a 3872x2592 image would have 10,036,224 pixels, or approximately 10 megapixels.

The pixel count alone is commonly presumed to indicate the resolution of a camera, but this simple figure of merit is a misconception. Other factors impact a sensor's resolution, including sensor size, lens quality, and the organization of the pixels (for example, a monochrome camera without a Bayer filter mosaic has a higher resolution than a typical color camera).

Where such other factors limit the resolution, a greater pixel count does not improve it, but may rather make the digital images inconveniently large and/or exacerbate image noise. Many digital compact cameras are criticized for having excessive pixels. Sensors can be so small that their 'buckets' can easily overfill; again, resolution of a sensor can become greater than the camera lens could possibly deliver.

Demanding high quality and resolution (e.g. for use in professional photography), this count is an object of manufacturer competition. As of August 2011, the highest resolution available on the market is 80.1 MP.[10]

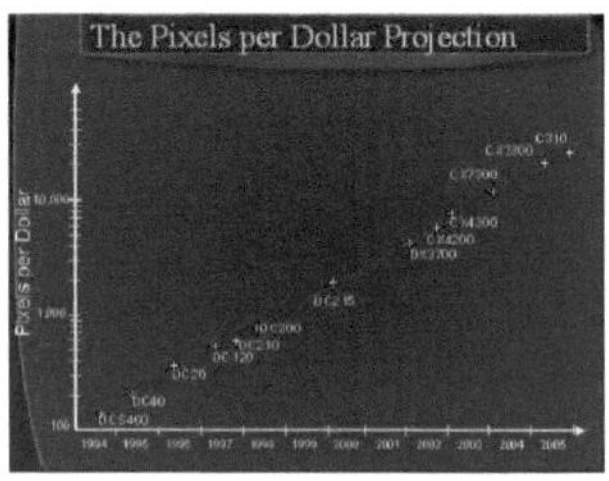

Australian recommended retail price of Kodak digital cameras.

As the technology has improved, costs have decreased dramatically. Counting the "pixels per dollar" as a basic measure of value for a digital camera, there has been a continuous and steady increase in the number of pixels each dollar buys in a new camera, in accord with the principles of Moore's Law. This predictability of camera prices was first presented in 1998 at the Australian PMA DIMA conference by Barry Hendy and since referred to as "Hendy's Law".[11]

Since only a few aspect ratios are commonly used (mainly 4:3 and 3:2), the number of sensor sizes that are useful is limited. Furthermore, sensor manufacturers do not produce every possible sensor size, but take incremental steps in sizes. For example, in 2007 the three largest sensors (in terms of pixel count) used by Canon were the 21.1, 17.9, and 16.6 megapixel CMOS sensors.

Methods of image capture

Since the first digital backs were introduced, there have been three main methods of capturing the image, each based on the hardware configuration of the sensor and color filters.

The first method is often called *single-shot*, in reference to the number of times the camera's sensor is exposed to the light passing through the camera lens. Single-shot capture systems use either one CCD with a Bayer filter mosaic, or three separate image sensors (one each for the primary additive colors red, green, and blue) which are exposed to the same image via a beam splitter.

At the heart of a digital camera is a CCD or a CMOS image sensor.

The second method is referred to as *multi-shot* because the sensor is exposed to the image in a sequence of three or more openings of the lens aperture. There are several methods of application of the multi-shot technique. The most common originally was to use a single image sensor with three filters (once again red, green and blue) passed in front of the sensor in sequence to obtain the additive color information. Another multiple shot method is called Microscanning. This technique utilizes a single CCD with a Bayer filter but actually moved the physical location of the sensor chip on the focus plane of the lens to "stitch" together a higher resolution image than the CCD would allow otherwise. A third version combined the two methods without a Bayer filter on the chip.

This digital camera is partly disassembled. The lens assembly (bottom right) is partially removed, but the sensor (top right) still captures a usable image, as seen on the LCD screen (bottom left).

The third method is called *scanning* because the sensor moves across the focal plane much like the sensor of a desktop scanner. Their *linear* or *tri-linear* sensors utilize only a single line of photosensors, or three lines for the three colors. In some cases, scanning is accomplished by moving the sensor e.g. when using Color co-site sampling or rotate the whole camera; a digital rotating line camera offers images of very high total resolution.

The choice of method for a given capture is determined largely by the subject matter. It is usually inappropriate to attempt to capture a subject that moves with anything but a single-shot system. However, the higher color fidelity and larger file sizes and resolutions available with multi-shot and scanning backs make them attractive for commercial photographers working with stationary subjects and large-format photographs.

Dramatic improvements in single-shot cameras and raw image file processing at the beginning of the 21st century made single shot, CCD-based cameras almost completely dominant, even in high-end commercial photography. CMOS-based single shot cameras remained somewhat common.

Filter mosaics, interpolation, and aliasing

The Bayer arrangement of color filters on the pixel array of an image sensor.

Most current consumer digital cameras use a Bayer filter mosaic in combination with an optical anti-aliasing filter to reduce the aliasing due to the reduced sampling of the different primary-color images. A demosaicing algorithm is used to interpolate color information to create a full array of RGB image data.

Cameras that use a beam-splitter single-shot 3CCD approach, three-filter multi-shot approach, Color co-site sampling or Foveon X3 sensor do not use anti-aliasing filters, nor demosaicing.

Firmware in the camera, or a software in a raw converter program such as Adobe Camera Raw, interprets the raw data from the sensor to obtain a full color image, because the RGB color model requires three intensity values for each pixel: one each for the red, green, and blue (other color models, when used, also require three or more values per pixel). A single sensor element cannot simultaneously record these three intensities, and so a color filter array (CFA) must be used to selectively filter a particular color for each pixel.

The Bayer filter pattern is a repeating 2×2 mosaic pattern of light filters, with green ones at opposite corners and red and blue in the other two positions. The high proportion of green takes advantage of properties of the human visual system, which determines brightness mostly from green and is far more sensitive to brightness than to hue or saturation. Sometimes a 4-color filter pattern is used, often involving two different hues of green. This provides potentially more accurate color, but requires a slightly more complicated interpolation process.

The color intensity values not captured for each pixel can be interpolated (or guessed) from the values of adjacent pixels which represent the color being calculated.

Sensor size and angle of view

Cameras with digital image sensors that are smaller than the typical 35mm film size has a smaller field or angle of view when used with a lens of the same focal length. This is because angle of view is a function of both focal length and the sensor or film size used.

If a sensor smaller than the full-frame 35mm film format is used, as in most digicams, then the field of view is cropped by the sensor to smaller than the 35mm full-frame format's field of view. This narrowing of the field of view is often described in terms of a *focal length multiplier* or crop factor, a factor by which a longer focal length lens would be needed to get the same field of view on a full-frame camera.

The result is geometrically similar to taking the image from the film camera and cutting it down (cropping) to the size of the sensor, ignoring various questions such as resolution. For moderately large DSLRs the crop factor may be in the range of 1.3-2 while smaller cameras use smaller sensors with a larger crop factor.

If the digital sensor has a higher or lower density of pixels per unit area than the film equivalent, then the amount of information captured differs correspondingly. While resolution can be estimated in pixels per unit area, the comparison is complex since most types of digital sensor record only a single colour at each pixel location, and different types of film have different effective resolutions. There are various trade-offs involved, since larger sensors are more expensive to manufacture and require larger lenses, while sensors with higher numbers of pixels per unit area are likely to suffer higher noise levels.

For these reasons, it is possible to obtain cheap digital cameras with sensor sizes much smaller than 35mm film, but with high pixel counts, that can still produce high-resolution images. Such cameras are usually supplied with lenses that would be classed as extremely wide angle on a 35mm camera, and that can also be smaller size and less expensive, since there is a smaller sensor to illuminate. For example, a camera with a 1/1.8" sensor has a 5.0x field of view crop, and so a hypothetical 5-50mm zoom lens produces images that look *similar* (again the differences mentioned above are important) to those produced by a 35mm film camera with a 25–250mm lens, while being much more compact than such a lens for a 35mm camera since the imaging circle is much smaller.

This can be useful if extra telephoto reach is desired, as a certain lens on an APS sensor produces an image equivalent to a significantly longer lens on a 35mm film camera shot at the same distance from the subject, the equivalent length of which depends on the camera's field of view crop. This is sometimes referred to as the focal length multiplier, but the focal length is a physical attribute of the lens and not the camera system itself. The disadvantage of this is that wide angle photography is made somewhat more difficult, as the smaller sensor effectively and undesirably reduces the captured field of view. Some methods of compensating for this or otherwise producing much wider digital photographs involve using a fisheye lens and "defishing" the image in post processing to simulate a rectilinear wide angle lens.

Full-frame digital SLRs, that is, those with sensor size matching a frame of 35mm film, include Canon 1Ds and 5D series, Kodak Pro DCS-14n, Nikon D3 line and Contax N Digital. There are very few digital cameras with sensors that can approach the resolution of larger-format film cameras, with the possible exception of the Mamiya ZD (22MP) and the Hasselblad H3D series of DSLRs (22 to 39 MP).

Common values for field of view crop in DSLRs include 1.3x for some Canon (APS-H) sensors, 1.5x for Sony APS-C sensors used by Nikon, Pentax and Konica Minolta and for Fujifilm sensors, 1.6 (APS-C) for most Canon sensors, ~1.7x for Sigma's Foveon sensors and 2x for Kodak and Panasonic 4/3" sensors currently used by Olympus and Panasonic. Crop factors for non-SLR consumer compact and bridge cameras are larger, frequently 4x or more.

Further information: Image sensor format

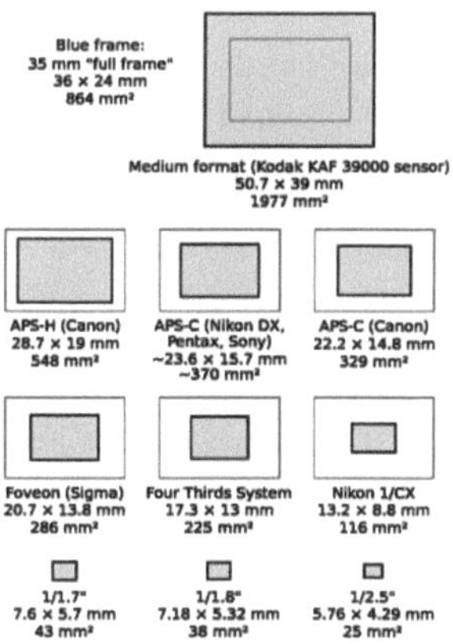

Relative sizes of sensors used in most current digital cameras.

Table of sensor sizes [12]			
Type	**Width (mm)**	**Height (mm)**	**Size (mm²)**
1/3.6"	4.00	3.00	12.0
1/3.2"	4.54	3.42	15.5
1/3"	4.80	3.60	17.3
1/2.7"	5.37	4.04	21.7
1/2.5"	5.76	4.29	24.7
1/2.3"	6.16	4.62	28.5
1/2"	6.40	4.80	30.7
1/1.8"	7.18	5.32	38.2
1/1.7"	7.60	5.70	43.3
2/3"	8.80	6.60	58.1
1"	12.8	9.6	123
4/3"	18.0	13.5	243
APS-C	25.1	16.7	419
35 mm	36	24	864
Back	48	36	1728

Connectivity

Transferring photos

Many digital cameras can connect directly to a computer to transfer data:

- Early cameras used the PC serial port. USB is now the most widely used method (most cameras are viewable as USB mass storage), though some have a FireWire port. Some cameras use USB PTP mode for connection instead of USB MSC; some offer both modes.
- Other cameras use wireless connections, via Bluetooth or IEEE 802.11 Wi-Fi, such as the Kodak EasyShare One.
- Cameraphones and some high-end stand-alone digital cameras also use cellular networks to connect for sharing images. The most common standard on cellular networks is the MMS Multimedia Messaging Service, commonly called "picture messaging". The second method with smartphones is to send a picture as an email attachment. Many cameraphones do not support email, so this is less common.

A common alternative is the use of a card reader which may be capable of reading several types of storage media, as well as high speed transfer of data to the computer. Use of a card reader also avoids draining the camera battery during the download process. An external card reader allows convenient direct access to the images on a collection of storage media. But if only one storage card is in use, moving it back and forth between the camera and the reader can be inconvenient. Many computers have a card reader built in, at least for SD cards.

Printing photos

Many modern cameras support the PictBridge standard, which allows them to send data directly to a PictBridge-capable computer printer without the need for a computer.

Wireless connectivity can also provide for printing photos without a cable connection.

Polaroid has introduced a printer integrated into its digital camera which creates a small, printed copy of a photo. This is reminiscent of the original instant camera, popularized by Polaroid in 1975.[13]

Displaying photos

Many digital cameras include a video output port. Usually sVideo, it sends a standard-definition video signal to a television, allowing the user to show one picture at a time. Buttons or menus on the camera allow the user to select the photo, advance from one to another, or automatically send a "slide show" to the TV.

HDMI has been adopted by many high-end digital camera makers, to show photos in their high-resolution quality on an HDTV.

In January 2008, Silicon Image announced a new technology for sending video from mobile devices to a television in digital form. MHL sends pictures as a video stream, up to 1080p resolution, and is compatible with HDMI.[14]

Some DVD recorders and television sets can read memory cards used in cameras; alternatively several types of flash card readers have TV output capability.

Modes

Many digital cameras have preset modes for different applications. Within the constraints of correct exposure various parameters can be changed, including exposure, aperture, focusing, light metering, white balance, and equivalent sensitivity. For example a portrait might use a wider aperture to render the background out of focus, and would seek out and focus on a human face rather than other image content.

Image data storage

Many camera phones and most separate digital cameras use memory cards having flash memory to store image data. The majority of cards for separate cameras are SD format; many are CompactFlash or other formats.

A CompactFlash (CF) card, one of many media types used to store digital photographs

Digital cameras have computers inside, hence have internal memory, and many cameras can use some of this internal memory for a limited capacity for pictures that can be transferred to or from the card or through the camera's connections.

A few cameras use some other form of removable storage such as Microdrives (very small hard disk drives), CD single (185 MB), and 3.5" floppy disks. Other unusual formats include:

- Onboard flash memory — Cheap cameras and cameras secondary to the device's main use (such as a camera phone)
- PC Card hard drives — early professional cameras (discontinued)
- Thermal printer — known only in one model of camera that printed images immediately rather than storing

Mini CD (left)

Microdrive (CF-II)

USB flash drive

3.5" floppy disks

Most manufacturers of digital cameras do not provide drivers and software to allow their cameras to work with Linux or other free software. Still, many cameras use the standard USB storage protocol, and are thus easily usable. Other cameras are supported by the gPhoto project.

File formats

The Joint Photography Experts Group standard (JPEG) is the most common file format for storing image data. Other file types include Tagged Image File Format (TIFF) and various Raw image formats.

Many cameras, especially high-end ones, support a raw image format. A raw image is the unprocessed set of pixel data directly from the camera's sensor, often saved in a proprietary format. Adobe Systems has released the DNG format, a royalty-free raw image format used by at least 10 camera manufacturers.

Raw files initially had to be processed in specialized image editing programs, but over time many mainstream editing programs, such as Google's Picasa, have added support for raw images. Rendering to standard images from raw sensor data allows more flexibility in making major adjustments without losing image quality or retaking the picture.

Formats for movies are AVI, DV, MPEG, MOV (often containing motion JPEG), WMV, and ASF (basically the same as WMV). Recent formats include MP4, which is based on the QuickTime format and uses newer compression algorithms to allow longer recording times in the same space.

Other formats that are used in cameras but not for pictures are the Design Rule for Camera Format (DCF), an ISO specification for the camera's internal file structure and naming, and Digital Print Order Format (DPOF), which dictates what order images are to be printed in and how many copies.

Most cameras include Exif data that provides metadata about the picture. Exif data may include aperture, exposure time, focal length, date and time taken, and location.

Batteries

Seller with cameras, memory cards & batteries at the Albuquerque International Balloon Fiesta, 2010.

Digital cameras have high power requirements, and over time have become smaller, resulting in an ongoing need to develop a battery small enough to fit in the camera and yet able to power it for a reasonable length of time.

Two broad types of batteries are in use for digital cameras.

Off-the-shelf

Off-the-shelf batteries may be single-use disposable or reusable rechargeable batteries. In either case they conform to an established off-the-shelf form factor, most commonly AA, CR2, or CR-V3 batteries, with AAA batteries in a handful of cameras. The CR2 and CR-V3 batteries are lithium based, and intended for single use. They are also commonly seen in camcorders. AA batteries are the most common; however, the non-rechargeable alkaline batteries supplied with low-end cameras are capable of powering most cameras for only a very short time. They may serve satisfactorily in cameras that are only occasionally used.

Consumers with more than an occasional need use AA Nickel metal hydride batteries (NiMH) instead, which provide adequate energy and are rechargeable. NIMH batteries do not provide as much energy per volume as lithium ion batteries, and they also tend to discharge when not used. For the same energy, a NiMH rechargeable battery takes up to twice the volume of a Li-on rechargeable battery, and is three to five times heavier, but only costs half as much. Rechargeable batteries are available in various ampere-hour (Ah) or milli-ampere-hour (mAh) ratings, which are approximately proportional to shots per charge.

Typically mid-range consumer models and some low end cameras use off-the-shelf batteries; only a very few DSLR cameras accept them (for example, Sigma SD10). Rechargeable RCR-V3 lithium-ion batteries are also available as an alternative to non-rechargeable CR-V3 batteries. Cameras, especially earlier ones made for AA-size batteries assumed that these would be of the non-rechargeable, preferably alkaline manganese type delivering 1.5 volts per cell. Rechargeable NiCd or NiMH cells only deliver 1.2 volts, which means that many such cameras will only operate for a short time or not at all even with new and newly charged 1.2 volt units. A portable ultra-high-endurance external power-supply for the shoulder bag to operate older 6 volt cameras can be made up of five 1.2 volt C-size cells which can be either NiCd or NiMH, with a cable and 4mm DC-plug.

Proprietary

The second type of battery for digital cameras is proprietary battery formats. These are built to a manufacturer's custom specifications, and can be either aftermarket replacement parts or OEM. Almost all proprietary batteries are lithium ion. While they only accept a certain number of recharges before the battery life begins degrading (typically up to 500 cycles), they provide considerable performance for their size. A result is that at the two ends of the spectrum both high end professional cameras and low end consumer models tend to use lithium ion batteries.

Conversion of film cameras to digital

When digital cameras became common, a question many photographers asked was whether their film cameras could be converted to digital. The answer was yes and no. For the majority of 35 mm film cameras the answer is no, the reworking and cost would be too great, especially as lenses have been evolving as well as cameras. For most a conversion to digital, to give enough space for the electronics and allow a liquid crystal display to preview, would require removing the back of the camera and replacing it with a custom built digital unit.

Digital single-lens reflex camera

Many early professional SLR cameras, such as the Kodak DCS series, were developed from 35 mm film cameras. The technology of the time, however, meant that rather than being digital "backs" the bodies of these cameras were mounted on large, bulky digital units, often bigger than the camera portion itself. These were factory built cameras, however, not aftermarket conversions.

A notable exception is the Nikon E2, followed by Nikon E3, using additional optics to convert the 35mm format to a 2/3 CCD-sensor.

A few 35 mm cameras have had digital camera backs made by their manufacturer, Leica being a notable example. Medium format and large format cameras (those using film stock greater than 35 mm), have a low unit production, and typical digital backs for them cost over $10,000. These cameras also tend to be highly modular, with handgrips, film backs, winders, and lenses available separately to fit various needs.

The very large sensor these backs use leads to enormous image sizes. For example Phase One's P45 39 MP image back creates a single TIFF image of size up to 224.6 MB, and even greater pixel counts are available. Medium format digitals such as this are geared more towards studio and portrait photography than their smaller DSLR counterparts; the ISO speed in particular tends to have a maximum of 400, versus 6400 for some DSLR cameras. (Canon EOS-1D Mark IV and Nikon D3S have ISO 12800 plus Hi-3 ISO 102400 with the Canon EOS-1Dx's ISO of 204800)

Digital camera backs

In the industrial and high-end professional photography market, some camera systems utilize modular (removable) image sensors. For example, some medium format SLR cameras, such as the Mamiya 645D series, allow installation of either a digital camera back or a traditional photographic film back.

- Area array
  - CCD
  - CMOS
- Linear array
  - CCD (monochrome)
  - 3-strip CCD with color filters

Linear array cameras are also called scan backs.

- Single-shot
- Multi-shot (three-shot, usually)

Most earlier digital camera backs used linear array sensors. The linear array sensor acts like its counterpart in a flatbed image scanner by moving vertically to digitize the image. Many early such cameras only capture grayscale images. Color photography requires three separate scans, and a mechanical assembly to cycle a primary color filter in front of the sensor. These are called multi-shot backs. The entire scanning process requires relatively long expsoure times, in the range of seconds or even minutes. Due to this relatively long exposure-time, scanning and mutli-shot backs are generally limited to studio applications, where all aspects of the photographic scene are under the photographer's control.

Some other camera backs use CCD arrays similar to typical cameras. These are called single-shot backs.

Since it is much easier to manufacture a high-quality linear CCD array with only thousands of pixels than a CCD matrix with millions, very high resolution linear CCD camera backs were available much earlier than their CCD matrix counterparts. For example, you could buy an (albeit expensive) camera back with over 7,000 pixel horizontal resolution in the mid-1990s. However, as of 2004, it is still difficult to buy a comparable CCD matrix camera of the same resolution. Rotating line cameras, with about 10,000 color pixels in its sensor line, are able, as of 2005, to capture about 120,000 lines during one full 360 degree rotation, thereby creating a single digital image of 1,200 Megapixels.

Most modern digital camera backs use CCD or CMOS matrix sensors. The matrix sensor captures the entire image frame at once, instead of incrementing scanning the frame area through the prolonged exposure. For example, Phase One produces a 39 million pixel digital camera back with a 49.1 x 36.8 mm CCD in 2008. This CCD array is a little smaller than a frame of 120 film and much larger than a 35 mm frame (36 x 24 mm). In comparison, consumer digital cameras use arrays ranging from 36 x 24 mm (full frame on high end consumer DSLRs) to 7.2 x 5.3 mm (on point and shoot cameras) CMOS sensor.

At present, there are relatively few complete digital SLR cameras with sensors large enough to compete with the image detail offered by medium to large format film cameras. Phase One, Mamiya, and Hasselblad in 2011 manufacture medium format digital devices that can capture 30MP up to 80MP. The units tend to be quite large and expensive. Additionally, because of their high build quality and lack of moving parts, they tend to be quite long lasting and are prominent on the used market.[15]

See also

- Charge-coupled device

References

[1] Musgrove, Mike (2006-01-12). "Nikon Says It's Leaving Film-Camera Business" (http://www.washingtonpost.com/wp-dyn/content/article/2006/01/11/AR2006011102323.html). Washington Post. . Retrieved 2007-02-23.

[2] http://www.gadgetell.com/tech/comment/ces-2011-kodak-intros -easyshare-touch-mini-and-sport-digital-cameras/

[3] 3D Cameras http://www.sony.co.uk/hub/cyber-shot-3d-camera/3/1

[4] http://www.cameras.co.uk/specs/panasonic-dmc-ft3.cfm

[5] Digital SLR vs Compact SuperZoom Camera http://www.safari-guide.co.uk/photography-dslr-vs-compact-superzoom-camera.php

[6] DP Review (http://www.dpreview.com/news/1103/11030211olysz30mr.asp)

[7] Kevin J. O'Brien, New York Times, 2010 Nov 15 (http://www.nytimes.com/2010/11/15/technology/15iht-navigate.html) Smartphone Sales Taking Toll on G.P.S. Devices

[8] *The Visual Dictionary of Photography* (http://books.google.com/books?id=f7X5vYbUd0sC&pg=PA91&dq=Steven+Sasson&hl=en&ei=McMtTpX1M8e5tgeA1qnXAg&sa=X&oi=book_result&ct=result&resnum=3&ved=0CD4Q6AEwAg#v=onepage&q&f=false). AVA Publishing. .

[9] "Steven Sasson named to CE Hall of Fame" (http://www.letsgodigital.org/en/16859/ce-hall-of-fame/). Let's Go Digital. .

[10] http://www.photographyblog.com/news/phase_one_iq180_iq160_and_iq140

[11] Bogdan Solca (2007-01-08). "More on digital cameras" (http://news.softpedia.com/news/Digital-Lens-Flares-43923.shtml). *Softpedia*. .

[12] Bockaert, Vincent. "Sensor sizes" (http://www.dpreview.com/learn/?/Glossary/Camera_System/sensor_sizes_01.htm). Digital Photography Review. . Retrieved 2007-04-03.

[13] "Reinventing Instant Photography for the Digital Age" (http://thenewinstant.com/news-updates/Polaroid Introduces the Polaroid PoGoâ¢ Instant Digital Camera/). Polaroid. January 8, 2009. . Retrieved 2009-01-15.

[14] "Mobile High-Definition Link Technology Gives Consumers the Ability to Link Mobile Devices to HDTVs with Support for Audio and Video" (http://www.siliconimage.com/news/releasedetails.aspx?id=480). Silicon Image. January 7, 2008. . Retrieved 2009-01-15.

[15] Used Phase One Medium Format Digital Backs, Capture Integration (http://www.captureintegration.com/phase-one/phase-one-tech-specs/)

External links

- Howstuffworks Digital Cameras (http://products.howstuffworks.com/digital-cameras-make-photography-a-snap-article.htm)
- Year-by-year developments up to 1998 and images of several types of cameras (http://www.digicamhistory.com)
- Architecture of a still-motion-/picture digital camera (http://andrea.eric.free.fr/eric/static/en/Frame-RD-Activities-System-Analysis.html)
- Digital Camera in the Classroom (http://sutterworkshops.wikispaces.com/DigitalCameraInClassroom)
- The invention of the digital camera at Kodak (1975!) (http://pluggedin.kodak.com/post/?ID=687843)

AA battery

An **AA battery** is a standard size of battery. Batteries of this size are the most commonly used type of in portable electronic devices. An AA battery is composed of a single electrochemical cell. The exact terminal voltage and capacity of an AA size battery depends on the cell chemistry.

An AA size rechargeable cell

The AA battery size was standardized by the American National Standards Institute in 1947, but had been used in flashlights and electrical novelties for some time before formal standardization. Battery nomenclature gives different designations depending on cell size and chemistry. For example, a zinc–carbon (Leclanché) AA cell is designated "15" by ANSI, and R6 in the IEC system.

In 2011, AA batteries accounted for approximately 60% of alkaline primary (non-rechargeable) battery sales in the United States. In Japan, 58% of alkaline primary batteries sold were AA. In Switzerland, AA batteries totaled 55% in both primary and secondary(rechargeable) battery sales. [1] [2] [3]

	Zinc–carbon	**Alkaline**	**Li-FeS$_2$**	**NiCd**	**NiMH**	**NiZn**
IEC name	R6	LR6	FR6	KR6	HR6	?
ANSI/NEDA name	15D	15A	15LF	1.2K2	1.2H2	?
Capacity under 500mA constant drain	400-1000 mAh	1800-2600 mAh	2700-3400 mAh	600– 1000 mAh	2200–2900 mAh	1500-1800 mAh
Nominal voltage	1.50 V	1.50 V	1.50 V	1.25 V	1.25 V	1.65 V
Rechargeable	No	No	No[4]	Yes	Yes	Yes

Dimensions

An AA battery measures 51 mm (2.0 in) in length—50.1 mm (1.97 in) without the button terminal—and 13.5–14.5 mm (0.53–0.57 in) in diameter. [5]

Alkaline AA batteries have a mass of roughly 23 g (0.81 oz), lithium AA batteries have a mass around 15 g (0.53 oz), and rechargeable NiMH batteries around 31 g (1.1 oz).

D, C, **AA**, AAA, AAAA & 9-Volt batteries

Primary chemistry and capacity

Primary (non-rechargeable) zinc–carbon (Leclanché cell) AA batteries have around 400–900 milli amp-hours capacity, depending entirely on test conditions, duty cycle, and cut-off voltage selected. Zinc–carbon batteries are usually marketed as "general purpose" batteries. Zinc-chloride batteries store around 1000 to 1500 mAh are often sold as "heavy duty" or "super heavy duty". Alkaline batteries from 1700 mAh to 3000 mAh cost a little more than zinc-chloride batteries, but last proportionally longer.

Non-rechargeable lithium batteries are manufactured for devices that drain a large amount of power such as digital cameras, where their high cost is offset by longer running time between battery changes.

Rechargeable chemistry and capacity

Rechargeable Nickel–cadmium battery (NiCd or NiCad) AAs with a capacity of 500 – 1100 mAh are available, cost increasing with capacity. Nickel–metal hydride (NiMH) AAs are also available in various capacities of 1300 – 3000 mAh.

Low self-discharge NiMh batteries are also made in the AA size.

Rechargeable AA-sized batteries based on Li-ion chemistry have also been introduced. These batteries are rated at 3.6 volts and are incompatible with most AA-based devices.[6] AA lithiums have a relatively low internal resistance that effectively provides very high current if shorted. Equipment designed to rely on a battery's internal resistance to limit current may not tolerate higher inrush current with lithium cells.

A solar-powered charger for rechargeable AA batteries

Where space permits, a pair of AA batteries can sometimes be replaced by a single CR-V3 battery. [7]

Other common names

U7 (In Britain until the 1980s)
Pencil-sized
Penlight
Mignon
MN1500
MX1500
Type 316 (Soviet Union/Russia)
UM 3 (JIS)

#5 (China)
6135-99-052-0009 (NSN)(carbon-zinc)
6135-99-195-6708 (NSN)(alkaline)
Finger Battery - "Paluszek" (Poland)

See also

- List of battery sizes
- Battery nomenclature

References

[1] (http://www.inobat.ch/fileadmin/user_upload/pdf_09/Absatz_Statistik_2008.pdf) INOBAT 2008 statistics.
[2] Life Cycle Impacts of Alkaline Batteries with a Focus on End-of-Life - EPBA-EU (http://www.epbaeurope.net/documents/NEMA_alkalinelca2011.pdf)
[3] Monthly battery sales statistics - MoETI - March 2011 (http://www.baj.or.jp/e/statistics/02.php)
[4] http://data.energizer.com/PDFs/lithiuml91l92_appman.pdf
[5] "Rabbit Season" (http://people.msoe.edu/~tritt/trips/rabbitseason.html). People.msoe.edu. . Retrieved 2010-06-19.
[6] "AA and AAA Battery FAQs" (http://www.greenbatteries.com/aa-battery-faqs.html). Greenbatteries.com. . Retrieved 2010-06-19.
[7] "welcome to NEXcell" (http://www.nexcellglobal.com/RCRV3.htm). Nexcellglobal.com. . Retrieved 2010-06-19.

External links

- Detailed specifications for Energizer alkaline AA battery (http://data.energizer.com/PDFs/E91.pdf)
- Detailed specifications for Duracell alkaline AA battery (http://www1.duracell.com/oem/Pdf/new/MX1500_US_UL.pdf)
- Brand Neutral Drawing Of AA Alkaline battery based on ANSI Specifications (http://www.batteryholders.org/aa.pdf)
- Brand Neutral Drawing Of AA NiCad battery based on ANSI Specifications (http://www.batteryholders.org/AA-NiCd.pdf)
- Brand Neutral Drawing Of AA Lithium battery based on ANSI Specifications (http://www.batteryholders.org/AA-LITHIUM.pdf)
- Brand Neutral Drawing Of AA NiMH battery based on ANSI Specifications (http://www.batteryholders.org/AA-NiMH.pdf)
- Brand Neutral Drawing Of AA Rechargeable battery based on ANSI Specifications (http://www.batteryholders.org/AA-RECHARGE.pdf)

Zoom lens

A **zoom lens** is a mechanical assembly of lens elements for which the focal length (and thus angle of view) can be varied, as opposed to a fixed focal length (FFL) lens (see prime lens). They are commonly used with still, video, motion picture cameras, projectors, some binoculars, microscopes, telescopes, telescopic sights, and other optical instruments.

Nikkor 28-200 mm zoom lens, extended to 200 mm at left and collapsed to 28 mm focal length at right

A true zoom lens, also called a parfocal lens, is one that maintains focus when its focal length changes. A lens that loses focus during zooming is more properly called a varifocal lens.

Applications

A photograph taken with a zoom lens, in which the focal length was varied during the course of the exposure.

Zoom lenses are often described by the ratio of their longest to shortest focal lengths. For example, a zoom lens with focal lengths ranging from 100 mm to 400 mm may be described as a 4:1 or "4×" zoom. The term superzoom or hyperzoom is used to describe photographic zoom lenses with very large focal length factors, typically more than 4× and ranging up to 15× in SLR camera lenses and 36× in amateur digital cameras. This ratio can be as high as 100× in professional television cameras.[1] As of 2009, photographic zoom lenses beyond about 3× cannot generally produce imaging quality on par with prime lenses. Constant fast aperture zooms (usually *f*/2.8 or *f*/2.0) are typically restricted to this zoom range. Quality degradation is less perceptible when recording moving images at low resolution, which is why professional video and TV lenses are able to feature high zoom ratios. Digital photography can also accommodate algorithms that compensate for optical flaws, both within in-camera processors and post-production software.

Some photographic zoom lenses are long-focus lenses, with focal lengths longer than a normal lens, some are wide-angle lenses (wider than *normal*), and others cover a range from wide-angle to long-focus. Lenses in the latter group of zoom lenses, sometimes referred to as "normal" zooms, have displaced the fixed focal length lens as the popular one-lens selection on many contemporary cameras. The markings on these lenses usually say **W** and **T** for "Wide" and "Telephoto". Telephoto is designated because the longer focal length supplied by the negative diverging lens is longer than the overall lens assembly (the negative diverging lens acting as the "telephoto group").[2]

Some digital cameras allow cropping and enlarging of a captured image, in order to emulate the effect of a longer focal length zoom lens (narrower angle of view). This is commonly known as digital zoom and produces an image of lower optical resolution than optical zoom. Exactly the same effect can be obtained by using digital image processing software on a computer to crop the digital image and enlarge the cropped area. Many digital cameras have both, combining them by first using the optical, then the digital zoom.

In addition to its photographic use, the afocal part of a zoom lens can be used as a telescope of variable magnification to make an adjustable beam expander. This can be used, for example, to change the size of a laser beam so that the irradiance of the beam can be varied.

History

The Voigtländer Zoomar, 36–82 mm *f*/2.8

Early forms of zoom lenses were used in optical telescopes to provide continuous variation of the magnification of the image, and this was first reported in the proceedings of the Royal Society in 1834. Early patents for telephoto lenses also included movable lens elements which could be adjusted to change the overall focal length of the lens. Lenses of this kind are now called **varifocal lenses**, since when the focal length is changed, the position of the focal plane also moves, requiring refocusing of the lens after each change.

The first true *zoom* lens, which retained near-sharp focus while the effective focal length of the lens assembly was changed, was patented in 1902 by Clile C. Allen (U.S. Patent 696788 [3]). The first industrial production was the Bell and Howell Cooke "Varo" 40–120 mm lens for 35mm movie cameras introduced in 1932. The most impressive early TV Zoom lens was the VAROTAL III from Rank Taylor Hobson from UK built in 1953. The Kilfitt 36–82 mm/2.8 Zoomar introduced in 1959 was the first varifocal lens in regular production for still 35mm photography. The first modern film zoom lens was designed around 1950 by Roger Cuvillier, a French engineer working for SOM-Berthiot. It had an optical compensation zoom system. In 1956, Pierre Angenieux introduced the mechanical compensation system, enabling precise focus while zooming, in his 10x lens released in 1958. Angenieux received a 1964 technical award from the academy of motion pictures for the design of that 12-120 zoom lens.

Since then advances in optical design, particularly the use of computers for optical ray tracing, has made the design and construction of zoom lenses much easier, and they are now used widely in professional and amateur photography.

Design

There are many possible designs for zoom lenses, the most complex ones having upwards of thirty individual lens elements and multiple moving parts. Most, however, follow the same basic design. Generally they consist of a number of individual lenses that may be either fixed, or slide axially along the body of the lens. While the magnification of a zoom lens changes, it is necessary to compensate for any movement of the focal plane to keep the focused image sharp. This compensation may be done by mechanical means (moving the complete lens assembly while the magnification of the lens changes), or optically (arranging the position of the focal plane to vary as little as possible while the lens is zoomed).

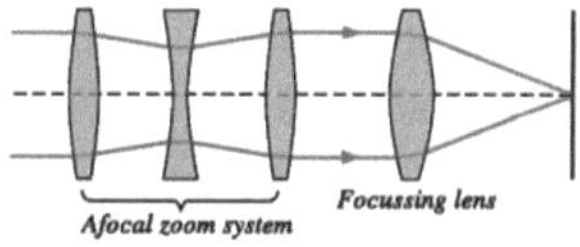

A simple zoom lens system. The three lenses of the afocal system are L_1, L_2, L_3 (from left). L_1 and L_2 can move to the left and right, changing the overall focal length of the system (see image below).

A simple scheme for a zoom lens divides the assembly into two parts: a focussing lens similar to a standard, fixed-focal-length photographic lens, preceded by an *afocal zoom system*, an arrangement of fixed and movable lens elements that does not focus the light, but alters the size of a beam of light travelling through it, and thus the overall magnification of the lens system.

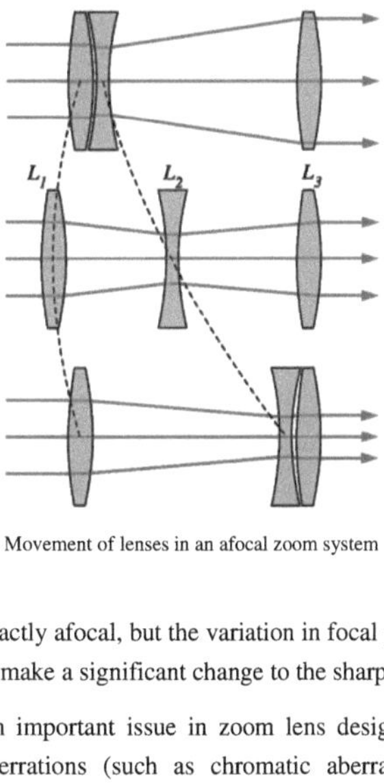

Movement of lenses in an afocal zoom system

In this simple optically compensated zoom lens, the afocal system consists of two positive (converging) lenses of equal focal length (lenses L_1 and L_3) with a negative (diverging) lens (L_2) between them, with an absolute focal length less than half that of the positive lenses. Lens L_3 is fixed, but lenses L_1 and L_2 can be moved axially, and do so in a fixed, non-linear relationship. This movement is usually performed by a complex arrangement of gears and cams in the lens housing, although some modern zoom lenses use computer-controlled servos to perform this positioning.

While the negative lens L_2 moves from the front to the back of the lens, the lens L_1 moves forward and then backward in a parabolic arc. In doing so, the overall angular magnification of the system varies, changing the effective focal length of the complete zoom lens. At each of the three points shown, the three-lens system is afocal (neither diverging or converging the light), and so does not alter the position of the focal plane of the lens. Between these points, the system is not exactly afocal, but the variation in focal plane position can be small enough (~±0.01 mm in a well-designed lens) not to make a significant change to the sharpness of the image.

An important issue in zoom lens design is the correction of optical aberrations (such as chromatic aberration, and in particular field curvature) across the whole operating range of the lens; this is considerably harder in a zoom lens than a fixed lens, which needs only to correct the aberrations for one focal length. This problem was a major reason for the slow uptake of zoom lenses, with early designs being considerably inferior to contemporary fixed lenses, and usable only with a narrow range of f-numbers. Modern optical design techniques have enabled the construction of zoom lenses with good aberration correction over widely variable focal lengths and apertures.

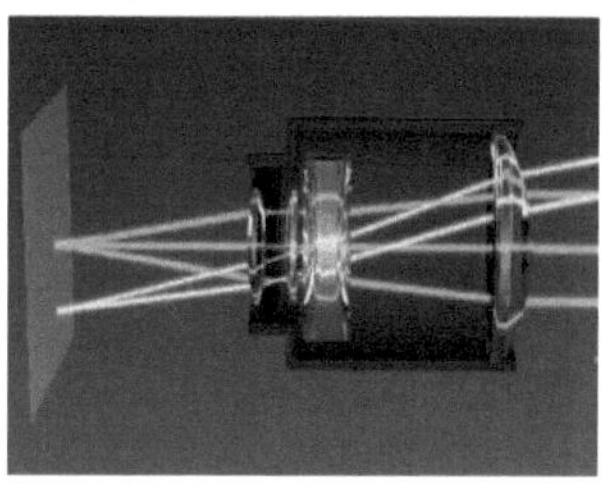

Simplified zoom lens in operation

Whereas lenses used in cinematography and video applications are required to maintain focus while the focal length is changed, there is no such requirement for still photography, or if a zoom lens is used as a projection lens. Since it is harder to construct a lens that does not change focus with the same image quality as one that does, the latter applications often have lenses that require refocusing once the focal length has changed (and thus strictly speaking are varifocal lenses, not zoom lenses). As most modern still cameras are autofocus, this is not a problem.

Designers of zoom lenses with large zoom ratios often trade one or more aberrations for higher image sharpness. For example, a greater degree of barrel and pincushion distortion is tolerated in lenses that span the focal length range from wide angle to telephoto with a focal ratio of 10x or more than would be acceptable in a fixed focal length lens or a zoom lens with a lower ratio. Although modern design methods have been continually reducing this problem, barrel distortion of greater than one percent is common in these large-ratio lenses. Another price paid is that at the extreme telephoto setting of the lens the effective focal length changes significantly while the lens is focused on closer objects. The apparent focal length can more than halve while the lens is focused from infinity to medium close-up. To a lesser degree, this effect is also seen in fixed focal length lenses that move internal lens elements, rather than the entire lens, to effect changes in magnification.

Varifocal lens

Many so-called "zoom" lenses, particularly in the case of fixed lens cameras, are actually *varifocal lenses*, which gives lens designers more flexibility in optical design trade-offs (focal length range, maximum aperture, size, weight, cost) than true parfocal zoom, and which is practical because of auto-focus, and because the camera processor can move the lens to compensate for the change in the position of the focal plane while changing magnification ("zooming"), making operation essentially the same as a true parfocal zoom.[4]

See also

- Pan tilt zoom camera (PTZ)
- Professional video camera
- Tripod (photography)
- View camera

By focal length:

- Prime lens
- Superzoom
- Telephoto lens
- Wide-angle lens

Focus attributes:

- Parfocal lens
- Varifocal lens

By system:

- List of Canon EF lenses
- Nikon F-mount
- Minolta AF

References

[1] "DIGISUPER 100 xs" (http://www.canon.com/bctv/products/digi100xs.html). Canon Broadcast Equipment. .

[2] Business and corporate aviation management: on demand air transportation By John J. Sheehan, page 521 (http://books.google.com/books?id=LbdefSMHICkC&pg=PA521&dq=zoom+lens+negative+become+telephoto&hl=en&ei=2EAKToW7CaTg0QHPpdmeAQ&sa=X&oi=book_result&ct=result&resnum=4&ved=0CDYQ6AEwAw#v=onepage&q=zoom lens negative become telephoto&f=false)

[3] http://www.google.com/patents?vid=696788

[4] Cavanagh, Roger (2003-05-29). "Parfrocal Lenses" (http://web.archive.org/web/20071007070643/http://www.rogercavanagh.com/helpinfo/30_parfocal.stm). Archived from the original (http://www.rogercavanagh.com/helpinfo/30_parfocal.stm) on 2007-10-07. . Retrieved 2007-11-18.

- Kingslake, R. (1960), "The development of the zoom lens". *Journal of the SMPTE* **69**, 534
- Clark, A.D. (1973), *Zoom Lenses, Monographs on Applied Optics No. 7.* Adam Hildger (London).
- Malacara, Daniel and Malacara, Zacarias (1994), *Handbook of Lens Design.* Marcel Dekker, Inc. ISBN 0-8247-9225-4
- " What is Inside a Zoom Lens? (http://www.adaptall-2.com/articles/InsideZoomLens/InsideZoomLens.html)", *Adaptall-2.com.*

External links

- Principe de fonctionnement du variateur de champ afocal(fr) (http://www.pierretoscani.com/echo_telezooms.html)

Focal length

The **focal length** of an optical system is a measure of how strongly the system converges (focuses) or diverges (defocuses) light. For an optical system in air, it is the distance over which initially collimated rays are brought to a focus. A system with a shorter focal length has greater optical power than one with a long focal length; that is, it bends the rays more strongly, bringing them to a focus in a shorter distance.

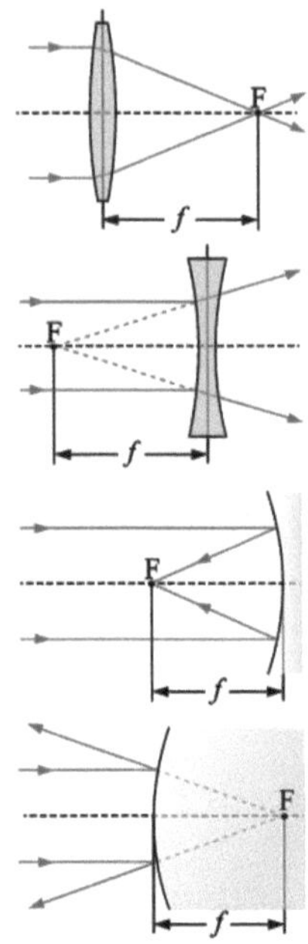

The focal point **F** and focal length *f* of a positive (convex) lens, a negative (concave) lens, a concave mirror, and a convex mirror.

In most photography and all telescopy, where the subject is essentially infinitely far away and magnification is achieved by moving the image plane away from the center of projection, longer focal length (lower optical power) leads to higher magnification and a narrower angle of view; conversely, shorter focal length or higher optical power is associated with a wider angle of view. On the other hand, in applications such as microscopy in which magnification is achieved by bringing the object close to the lens, a shorter focal length (higher optical power) leads to higher magnification because the subject can be brought closer to the center of projection.

Thin lens approximation

For a thin lens in air, the focal length is the distance from the center of the lens to the principal foci (or *focal points*) of the lens. For a converging lens (for example a convex lens), the focal length is positive, and is the distance at which a beam of collimated light will be focused to a single spot. For a diverging lens (for example a concave lens), the focal length is negative, and is the distance to the point from which a collimated beam appears to be diverging after passing through the lens.

General optical systems

For a *thick lens* (one which has a non-negligible thickness), or an imaging system consisting of several lenses and/or mirrors (e.g., a photographic lens or a telescope), the focal length is often called the **effective focal length** (EFL), to distinguish it from other commonly-used parameters:

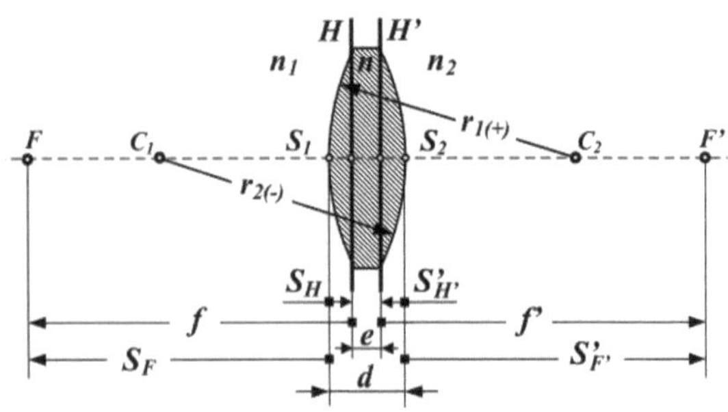

Thick lens diagram

- **Front focal length** (FFL) or **Front focal distance** (FFD) is the distance from the front focal point of the system to the vertex of the *first optical surface*.[1] [2]

- **Back focal length** (BFL) or **Back focal distance** (BFD) is the distance from the vertex of the *last optical surface* of the system to the rear focal point.[1] [2]

For an optical system in air, the effective focal length gives the distance from the front and rear principal planes to the corresponding focal points. If the surrounding medium is not air, then the distance is multiplied by the refractive index of the medium. Some authors call this distance the front (rear) focal length, distinguishing it from the front (rear) focal *distance*, defined above.[1]

In general, the focal length or EFL is the value that describes the ability of the optical system to focus light, and is the value used to calculate the magnification of the system. The other parameters are used in determining where an image will be formed for a given object position.

For the case of a lens of thickness d in air, and surfaces with radii of curvature R_1 and R_2, the effective focal length f is given by:

where n is the refractive index of the lens medium. The quantity $1/f$ is also known as the optical power of the lens.

The corresponding front focal distance is:

and the back focal distance:

In the sign convention used here, the value of R_1 will be positive if the first lens surface is convex, and negative if it is concave. The value of R_2 is positive if the second surface is concave, and negative if convex. Note that sign conventions vary between different authors, which results in different forms of these equations depending on the convention used.

For a spherically curved mirror in air, the magnitude of the focal length is equal to the radius of curvature of the mirror divided by two. The focal length is positive for a concave mirror, and negative for a convex mirror. In the sign convention used in optical design, a concave mirror has negative radius of curvature, so

,

where is the radius of curvature of the mirror's surface.

See Radius of curvature (optics) for more information on the sign convention for radius of curvature used here.

In photography

28 mm lens

50 mm lens

70 mm lens

210 mm lens

Camera lens focal lengths are usually specified in millimetres (mm), but some older lenses are marked in centimetres (cm) or inches.

When a photographic lens is set to "infinity", its rear nodal point is separated from the sensor or film, at the focal plane, by the lens's focal length. Objects far away from the camera then produce sharp images on the sensor or film, which is also at the image plane.

To render closer objects in sharp focus, the lens must be adjusted to increase the distance between the rear nodal point and the film, to put the film at the image plane. The focal length ( ), the distance from the front nodal point to the object to photograph (), and the distance from the rear nodal point to the image plane () are then related by:

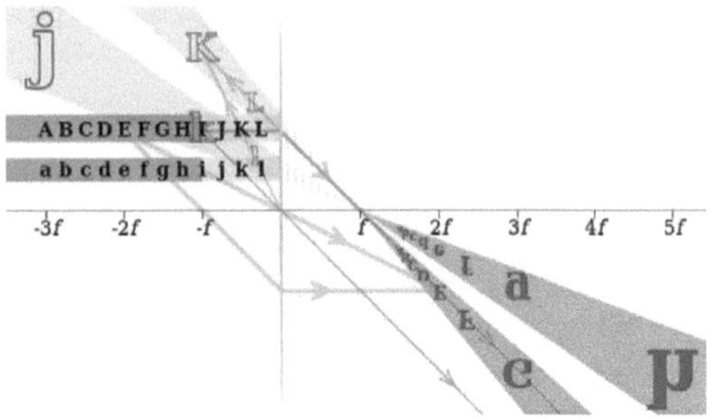

Images of black letters in a thin convex lens of focal length *f* are shown in red. Selected rays are shown for letters **E**, **I** and **K** in blue, green and orange, respectively. Note that **E** (at 2*f*) has an equal-size, real and inverted image; **I** (at *f*) has its image at infinity; and **K** (at *f*/2) has a double-size, virtual and upright image.

.

As is decreased, must be increased. For example, consider a normal lens for a 35 mm camera with a focal length of . To focus a distant object (), the rear nodal point of the lens must be located a distance from the image plane. To focus an object 1 m away (), the lens must be moved 2.6 mm further away from the image plane, to .

The focal length of a lens determines the magnification at which it images distant objects. It is equal to the distance between the image plane and a pinhole that images distant objects the same size as the lens in question. For rectilinear lenses (that is, with no image distortion), the imaging of distant objects is well modeled as a pinhole camera model.[3] This model leads to the simple geometric model that photographers use for computing the angle of view of a camera; in this case, the angle of view depends only on the ratio of focal length to film size. In general, the angle of view depends also on the distortion.[4]

A lens with a focal length about equal to the diagonal size of the film or sensor format is known as a normal lens; its angle of view is similar to the angle subtended by a large-enough print viewed at a typical viewing distance of the print diagonal, which therefore yields a normal perspective when viewing the print;[5] this angle of view is about 53 degrees diagonally. For full-frame 35mm-format cameras, the diagonal is 43 mm and a typical "normal" lens has a 50 mm focal length. A lens with a focal length shorter than normal is often referred to as a wide-angle lens (typically 35 mm and less, for 35mm-format cameras), while a lens significantly longer than normal may be referred to as a telephoto lens (typically 85 mm and more, for 35mm-format cameras). Technically long focal length lenses are only "telephoto" if the focal length is longer than the physical length of the lens, but the term is often used to describe any long focal length lens.

Due to the popularity of the 35 mm standard, camera–lens combinations are often described in terms of their 35 mm equivalent focal length, that is, the focal length of a lens that would have the same angle of view, or field of view, if used on a full-frame 35 mm camera. Use of a 35 mm equivalent focal length is particularly common with digital cameras, which often use sensors smaller than 35 mm film, and so require correspondingly shorter focal lengths to achieve a given angle of view, by a factor known as the crop factor.

See also

- Depth of field
- f-number or focal ratio
- Dioptre
- Focus (optics)

References

[1] John E. Greivenkamp (2004). *Field Guide to Geometrical Optics* (http://books.google.com/?id=1YfZNWZAwCAC&pg=PA6&dq="focal+length"+"rear+principal"+intitle:Geometrical#v=onepage&q="focal length" "rear principal" intitle:Geometrical&f=false). SPIE Press. p. 6–9. ISBN 9780819452948. .

[2] Hecht, Eugene (1987). *Optics* (2nd ed.). Addison Wesley. pp. 148–9. ISBN 0-201-11609-X.

[3] Jeffrey Charles (2000). *Practical astrophotography* (http://books.google.com/books?id=KuQErr3OlfOC&pg=PA63). Springer. p. 63–66. ISBN 9781852330231. .

[4] Leslie Stroebel and Richard D. Zakia (1993). *The Focal encyclopedia of photography* (http://books.google.com/books?id=CU7-2ZLGFpYC&pg=PA27) (3rd ed.). Focal Press. p. 27. ISBN 9780240514178. .

[5] Leslie D. Stroebel (1999). *View Camera Technique* (http://books.google.com/?id=71zxDuunAvMC&pg=PA136&dq=appear-normal+focal-length-lens+print-size+diagonal+viewer+distance). Focal Press. p. 135–138. ISBN 9780240803456. .

USB mass-storage device class

The **USB mass storage device class**, otherwise known as **USB MSC** or **UMS**, is a protocol that allows a Universal Serial Bus (USB) device to become accessible to a host computing device, to enable file transfers between the two. To the host device, the USB device appears similar to an external hard drive, enabling drag-and-drop file transfers.

The USB mass storage device class comprises a set of computing communications protocols defined by the USB Implementers Forum that run on the Universal Serial Bus. The standard provides an interface to a variety of storage devices.

Some of the devices that are connected to computers via this standard include:

- external magnetic hard drives
- external optical drives, including CD and DVD reader and writer drives
- portable flash memory devices
- adapters bridging between standard flash memory cards and USB connections
- digital cameras
- various digital audio players and portable media players
- card readers
- PDAs
- mobile phones

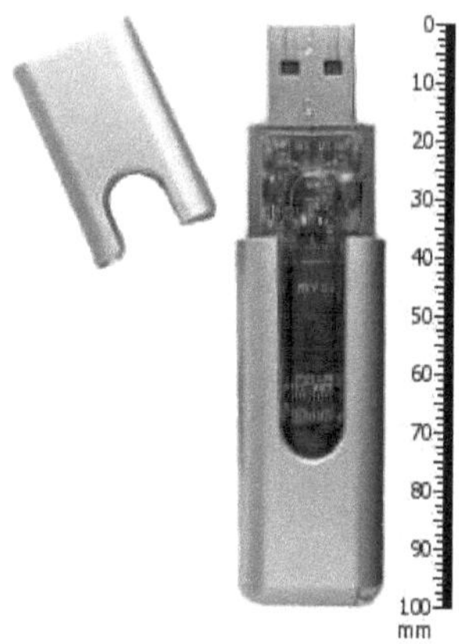

An USB flash drive like this one will typically implement the USB mass storage device class.

Devices which support this standard are referred to as **MSC** (Mass Storage Class) devices. While MSC is the official abbreviation, **UMS** (Universal Mass Storage) has become common in online jargon.

Operating system support

Most current mainstream operating systems include support for USB mass storage devices, although support on older systems is usually available through patches.

Microsoft Windows

Microsoft Windows natively supports MSC since Windows 2000 (Windows NT5).

There is no native (supplied by Microsoft) support for USB in Windows versions prior to Windows 95 and Windows NT 4.0, or MS-DOS. Windows 95 OSR2.1, an update to the operating system, featured very limited support for USB. During that time, no generic USB mass storage driver was produced by Microsoft, even for Windows 98. This meant that a device-specific driver was needed for each type of USB storage device.

Windows 2000 has support via a generic driver for mass storage devices, however a device-specific driver was required for rare devices not conforming the USB MSC specification). Windows Me and all later Windows versions also include support. Today, third party generic drivers which support USB flash drives are available freely even for Windows 98 SE and Windows NT 4.0.

Windows Mobile supports accessing most USB mass storage devices formatted with FAT on devices with USB Host. However, portable devices typically cannot provide enough power for disk enclosures containing hard drives (a 2.5" hard drive typically requires the maximum 2.5 W provided by the USB specification) without a self-powered USB hub. On the other way around, Windows Mobile devices cannot show their file systems as a mass storage device unless the device implementer explicitly decides to add such functionality. However, third party applications exist to add MSC emulation to most WM devices (commercial "Softick CardExport" and free "WM5torage"). Generally only memory cards can be exported, and not internal storage memory, due to the complications with filesystems outlined in the Device access section.

Malware and inherent vulnerability

In the past Windows's AutoRun feature worked indiscriminately on any removable media, allowing USB storage devices to become the infection entryway for computer viruses. As the FAT file system, often used on USB storage for its simplicity and wide compatibility, has few access control features, a user has no convenient way to protect his or her USB drives from infection after inserting into an untrusted computer, unless the device has a hardware read-only switch. To address this issue, Microsoft announced it would be limiting the AutoRun functionality to CD/DVD drives only starting in Windows 7 and future updates to previous versions of Windows.[1]

Mac OS

Apple Computer's Mac OS 9 and Mac OS X support USB mass storage natively. (Mac OS 8.5.1, an older OS, supported USB mass storage through an optional driver.)

Note : Some external DVD drives that are currently sold on the market require Mac OS X 10.4.6 or later. Thus, it seems that prior versions of Mac OS may not be 100 % fully compatible with the USB mass-storage device class (or at least not with its latest versions)

Unix-like

The Linux kernel has supported USB mass storage devices since kernel version 2.4 (year 2001), and a backport to kernel 2.2.18[2] has also been made. In Linux beyond generic driver for all devices conforming to USB mass storage device class, there are quirks, bug-fixes and additional functionalities for various devices, controllers and vendors to enable more functions (like ATA command pass-through for ATA-USB bridges, useful for S.M.A.R.T. or temperature monitoring, controlling spin-up and spin-down of a disks in such hard drives and other options). This also includes most Android based devices, as it is using Linux kernel.

Solaris also supports devices since the 2.8 release and up (year 1998), NetBSD since the 1.5 release and up (year 2000), FreeBSD since the 4.0 release and up (year 2000), and OpenBSD since the 2.7 release and up (year 2000). Digital UNIX (later known as Tru64 UNIX), supports USB and USB mass storage devices since version 4.0E (year 1998). AIX supports USB mass storage devices since the 5.3 T9 and 6.1 T3 releases, but it is not well supported and lacks some features (like partitioning and using it as general block device in system)[3] .

DOS

DOS has no generic support for USB but there are external drivers available which support USB mass storage devices. These are Duse, USBASPI and DOSUSB.

FreeDOS supports USB mass storage as ASPI devices.

Amiga

Amiga supports USB mass storage devices through Poseidon, a USB stack which has become a de facto standard. It supports various USB device types through a modular system of Hardware Independent Device Driver (HIDD) classes. Poseidon is used in AmigaOS, MorphOS and AROS. The supposedly final AmigaOS version of the Poseidon stack was released OEM licensed with the Deneb USB card in May 2008. AmigaOS 4.0 has its own USB stack called Sirion, though it can still use Poseidon. A new USB stack called ANAIIS (Another Native Amiga IO Interface Stack) is available for *all* Amiga platforms with Highway or Subway hardware.

The FAT16 and FAT32 file systems are supported by the FAT95 file system driver.

In September 2009, Poseidon was released as open source.

Game consoles

The Xbox 360 and PlayStation 3 support most mass storage devices for data transfer of media such as pictures and music. The Xbox 360 (as of April 2010) also allows using a mass storage device for saved games[4] , and the PS3 allows transferring saves between devices on a mass storage device.

Graphing calculators

Independent developers have released drivers for the TI-84 Plus and TI-84 Plus Silver Edition in order to access USB mass storage devices.[5] In these calculators, the usb8x driver supports the msd8x user interface application.

Device access

The USB mass storage specification does not require any particular file system to be used on conforming devices. Instead, it provides a simple interface to read and write sectors of data—much like the low-level interface used to access any hard drive—using the "SCSI transparent command set." Operating systems may treat the USB drive like a hard drive, and can format it with any file system they like. Creating multiple partitions is also possible.

A USB Card reader, such as this one, will typically implement the USB mass storage device class.

Because of its ubiquity and relative simplicity, the most common file system on embedded devices such as USB flash drives, cameras, or digital audio players is Microsoft's FAT or FAT32 file system with (optional) support for long names. Large USB-based hard disks may come formatted with NTFS, which is much less supported outside Microsoft Windows. However, a keydrive or any other device may be formatted using another filesystem (for example HFS Plus on an Apple Macintosh, or Ext2 under Linux, or Unix File System under Solaris or BSD). Of course, this choice may limit other operating systems' ability to access the contents of the device. Another storage options like LVM, partition tables and software encryption are also possible, depending on operating system.

In cameras, MP3 players, and similar gadgets which must access the file system independently from an external host, the FAT32 filesystem is typically preferred by device manufacturers. All such devices stops using file system (unmount) before making it available to host operating system, to prevent file system corruption or other damages (theoretically it is possible to use it by both device and host, by using read-only mode or by using so called cluster-aware file system, however non of them are used in any of existing commodity devices, and have limited usage even in theory, especially on embedded devices). Some devices have also switch or option in menu, which allows using them in read-only mode (write protection). Often used to share files to multiple computers, without risk of virus infection from any of them.

There is variety of partitioning schemes used by preformated devices. There are two main schemes used by vendors. First puts file system (most commonly FAT32) directly on the device without any partitions, effectively making it start from sector 0, without any additional boot sectors, headers or partitions. Second one, uses DOS partition table (and MBR code), with single (first) partition spanning entire device. This partition is often aligned to some high power of 2 of sectors (like 1 or 2MB), which is also common and important in solid state drives for best performance and durability. Some devices with embedded storage, acting like USB mass storage device (like MP3 players with USB port), will report that file system is damaged or missing, if formatted using different file system than FAT32 (or any other originally used there). However most devices which uses partition by default can be repartitioned (by shrinking size of first partition and file system) to have more partitions. Such devices will use first partition for its own operations, however after plugging into host system all partitions will be available.

There are also devices which are connected using single USB port, but act as multiple USB devices, one of which is a USB mass storage device. This is often used to simplify distribution and access to drivers and documentation, mostly for Microsoft Windows and Mac OS X operating system. Such drivers are necessary to make full use of the device (usually because the device does not fit any of the standard USB classes, or has additional functionality). An embedded USB mass storage device makes it possible to install additional drivers without the need for any CD-ROM disks, floppies or Internet access to vendor web site. This becomes more important as many modern systems are supplied without any optical or floppy drives. Often Internet access is not available, for example, because the device is used to provide network access - such as Wireless, GSM or Ethernet cards. The embedded USB mass storage is commonly made permanently read-only by the vendor, preventing accidental corruption and the use of it for other purposes (it can be sometimes updated using proprietary protocols when doing device's firmware upgrade). There are multiple other advantages of this way of distribution: decreased costs, simplified installation and fact that user cannot forget the drivers (a problem when using the device or selling the device).

Complications of the mass-storage device class

Hard drive-based devices

Many modern hard drives support additional advanced commands, such as Native Command Queuing, which may increase performance, Secure Erase, which allows all the data on the drive to be securely erased, and S.M.A.R.T., which allows a computer to measure various indicators of drive reliability. These exist as extensions to the basic low-level command sets used by hard drives, such as SCSI, Serial ATA, or Parallel ATA.

These features do not work when hard drives are encapsulated in some disk enclosure supporting the USB mass storage interface. Some USB mass storage interfaces provides a *generic* interface which only provides basic read/write commands, as outlined above. This works perfectly well for basic data transfer using hard drive-based devices, however it means that there is no simple way to send advanced, device-specific commands to USB mass storage devices (although devices may create their own communication protocols over the USB-standard "control interface").

The new eSATA standard for external drives promises to address this issue, since it extends the internal SATA bus to external disks without any intermediate translation layer.

See also

- Disk encryption software
- Media Transfer Protocol
- Picture Transfer Protocol
- USB flash drive

References

[1] http://blogs.technet.com/msrc/archive/2009/04/28/changes-in-windows-to-meet-changes-in-threat-landscape.aspx
[2] http://lxr.linux.no/#linux-old+v2.2.18/drivers/usb/usb-storage.c
[3] http://eserver.livejournal.com/27707.html
[4] http://majornelson.com/archive/2010/03/26/USB-Memory-Support-for-the-Xbox-360-coming-April-6th.aspx
[5] http://wikiti.brandonw.net/index.php?title=83Plus:Software:usb8x/Asm_Interface/MSD

External links

- Mass Storage device class specification (http://www.usb.org/developers/devclass_docs/usb-msc-overview-1.3b.pdf) — on the site of the USB Implementers Forum.
- Bootability specification Mass Storage bootability specification (http://www.usb.org/developers/devclass_docs/usb_msc_boot_1.0.pdf) — describes how bootable USB Mass Storage devices should work.
- "USB Mass Storage BOT" (http://www.usb.org/developers/devclass_docs/usbmassbulk_10.pdf) is a link that often rots, lately found buried at "USB" (http://usb.org) -> Developers -> Documents -> Class Specs -> Approved -> Mass Storage -> "Mass Storage Bulk Only 1.0".
- USB Mass Storage Device source code in FreeBSD (http://svn.freebsd.org/viewvc/base/head/sys/dev/usb/storage/)
- USB Mass Storage source code in Linux (http://www.kernel.org/git/?p=linux/kernel/git/torvalds/linux-2.6.git;a=tree;f=drivers/usb/storage)

USB

Original logo	
Type	Bus
Production history	
Designer	Compaq, DEC, IBM, Intel, Microsoft, NEC and Nortel
Designed	1994
Manufacturer	Intel, Compaq, Microsoft, NEC, Digital Equipment Corporation, IBM, Nortel
Superseded	Serial port, parallel port, game port, Apple Desktop Bus, PS/2 connector
General specifications	
Length	5 metres (maximum)
Width	11.5 mm (A-plug), 8.45 mm (B-plug),
Height	4.5 mm (A-plug), 7.78 mm (B-plug, pre-v3.0)
Hot pluggable	Yes
External	Yes
Cable	4 wires; 8 in USB 3.0
Pins	4 (1 supply, 2 data, 1 ground); 8 in USB 3.0 (additional 4 for SuperSpeed technology)
Connector	Unique
Electrical	
Signal	5 volt DC
Max. voltage	5 V(±5%)
Max. current	500–900 mA @ 5 V (depending on version)
Data	
Data signal	Packet data, defined by specifications
Width	1 bit
Bitrate	1.5/12/480/4,000 Mbit/s (depending on version)
Max. devices	127
Protocol	Serial
Pin out	

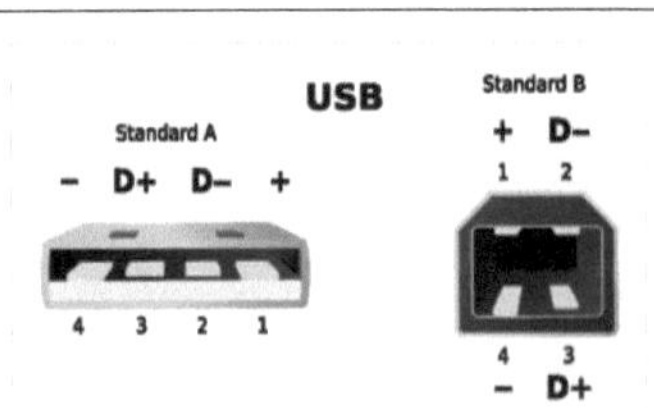

The standard USB A plug (left) and B plug (right)

Pin 1		V_{CC} (+5 V)	
Pin 2		Data-	

Pin 3		Data+	
Pin 4		Ground	

USB (**Universal Serial Bus**[1]) is an industry standard developed in the mid-1990s that defines the cables, connectors and protocols used in a bus for connection, communication and power supply between computers and electronic devices.

USB was designed to standardize the connection of computer peripherals, such as keyboards, pointing devices, digital cameras, printers, portable media players, disk drives and network adapters to personal computers, both to communicate and to supply electric power. It has become commonplace on other devices, such as smartphones, PDAs and video game consoles. USB has effectively replaced a variety of earlier interfaces, such as serial and parallel ports, as well as separate power chargers for portable devices.

As of 2008, approximately 6 billion USB devices had been sold in total, and about 2 billion were being sold each year.[2]

History

The basic USB *trident* logo; each released version has a specific logo variant

A group of seven companies began development on USB in 1994: Compaq, DEC, IBM, Intel, Microsoft, NEC and Nortel. The goal was to make it fundamentally easier to connect external devices to PCs by replacing the multitude of connectors at the back of PCs, addressing the usability issues of existing interfaces, and simplifying software configuration of all devices connected to USB, as well as permitting greater data rates for external devices. The first silicon for USB was made by Intel in 1995.[3]

The original USB 1.0 specification, which was introduced in January 1996, defined data transfer rates of 1.5 Mbit/s "Low Speed" and 12 Mbit/s "Full Speed".[3] The first widely used version of USB was 1.1, which was released in September 1998. The 12 Mbit/s data rate was intended for higher-speed devices such as disk drives, and the lower 1.5 Mbit/s rate for low data rate devices such as joysticks.[4]

A USB Standard Type A plug, the most common USB plug

The USB 2.0 specification was released in April 2000 and was ratified by the USB Implementers Forum (USB-IF) at the end of 2001. Hewlett-Packard, Intel, Lucent Technologies (now Alcatel-Lucent), NEC and Philips jointly led the initiative to develop a higher data transfer rate, with the resulting specification achieving 480 Mbit/s, a fortyfold increase over the original USB 1.1 specification.

The USB 3.0 specification was published on 12 November 2008. Its main goals were to increase the data transfer rate (up to 5 Gbit/s), to decrease power consumption, to increase power output, and to be backwards-compatible with USB 2.0.[5] USB 3.0 includes a new, higher speed bus called SuperSpeed in parallel with the USB 2.0 bus.[6] For this reason, the new version is also called SuperSpeed.[7] The first USB 3.0 equipped devices were presented in January 2010.[7] [8]

Version history

Prereleases

- *USB 0.7*: Released in November 1994.
- *USB 0.8*: Released in December 1994.
- *USB 0.9*: Released in April 1995.
- *USB 0.99*: Released in August 1995.
- *USB 1.0 Release Candidate*: Released in November 1995.

A PCI USB 2.0 card for a computer motherboard

USB 1.0

- *USB 1.0*: Released in January 1996.
  Specified data rates of *1.5 Mbit/s* (*Low-Bandwidth*) and *12 Mbit/s* (*Full-Bandwidth*). Does not allow for extension cables or pass-through monitors (due to timing and power limitations). Few such devices actually made it to market.
- *USB 1.1*: Released in August 1998.
  Fixed problems identified in 1.0, mostly relating to hubs. Earliest revision to be widely adopted.

USB 2.0

- *USB 2.0*: Released in April 2000.
  Added higher maximum bandwidth of *480 Mbit/s* (60 MB/s) (now called *"Hi-Speed"*). Further modifications to the USB specification have been done via Engineering Change Notices (ECN). The most important of these ECNs are included into the USB 2.0 specification package available from USB.org [9]:

The Hi-Speed USB Logo

  - *Mini-A and Mini-B Connector ECN*: Released in October 2000.
    Specifications for Mini-A and B plug and receptacle. Also receptacle that accepts both plugs for On-The-Go. These should not be confused with Micro-B plug and receptacle.
  - *Errata as of December 2000*: Released in December 2000.
  - *Pull-up/Pull-down Resistors ECN*: Released in May 2002.
  - *Errata as of May 2002*: Released in May 2002.
  - *Interface Associations ECN*: Released in May 2003.
    New standard descriptor was added that allows multiple interfaces to be associated with a single device function.
  - *Rounded Chamfer ECN*: Released in October 2003.
    A recommended, compatible change to Mini-B plugs that results in longer lasting connectors.
  - *Unicode ECN*: Released in February 2005.
    This ECN specifies that strings are encoded using UTF-16LE. USB 2.0 did specify that Unicode is to be used but it did not specify the encoding.
  - *Inter-Chip USB Supplement*: Released in March 2006.
  - *On-The-Go Supplement 1.3*: Released in December 2006.
    USB On-The-Go makes it possible for two USB devices to communicate with each other without requiring a separate USB host. In practice, one of the USB devices acts as a host for the other device.

- *Battery Charging Specification 1.1*: Released in March 2007 (Updated 15 Apr 2009).
  Adds support for dedicated chargers (power supplies with USB connectors), host chargers (USB hosts that can act as chargers) and the No Dead Battery provision which allows devices to temporarily draw 100 mA current after they have been attached. If a USB device is connected to dedicated charger, maximum current drawn by the device may be as high as 1.8 A. (Note that this document is not distributed with USB 2.0 specification package only USB 3.0 and USB On-The-Go.)
- *Micro-USB Cables and Connectors Specification 1.01*: Released in April 2007.
- *Link Power Management Addendum ECN*: Released in July 2007.
  This adds a new power state between enabled and suspended states. Device in this state is not required to reduce its power consumption. However, switching between enabled and sleep states is much faster than switching between enabled and suspended states, which allows devices to sleep while idle.

USB 3.0

The Super-Speed USB Logo

USB 3.0 was released in November 2008. The standard specifies a maximum transmission speed of up to 5 Gbit/s (640 MB/s), which is over 10 times faster than USB 2.0 (480 Mbit/s, or 60 MB/s), although this speed is typically only achieved using powerful professional grade or developmental equipment. USB 3.0 reduces the time required for data transmission, reduces power consumption, and is backward compatible with USB 2.0. The USB 3.0 Promoter Group announced on 17 November 2008 that the specification of version 3.0 had been completed and had made the transition to the USB Implementers Forum (USB-IF), the managing body of USB specifications.[10] This move effectively opened the specification to hardware developers for implementation in future products. A new feature is the "SuperSpeed" bus, which provides a fourth transfer mode at 5.0 Gbit/s. The raw throughput is 4 Gbit/s (using 8b/10b encoding), and the specification considers it reasonable to achieve around 3.2 Gbit/s (0.4 GB/s or 400 MB/s), increasing as hardware advances in the future take hold.

System design

A USB system has an asymmetric design, consisting of a host, a multitude of downstream USB ports, and multiple peripheral devices connected in a tiered-star topology. Additional USB hubs may be included in the tiers, allowing branching into a tree structure with up to five tier levels. A USB host may have multiple host controllers and each host controller may provide one or more USB ports. Up to 127 devices, including hub devices if present, may be connected to a single host controller.[11] [12]

USB devices are linked in series through *hubs*. There always exists one hub known as the root hub, which is built into the host controller.

A physical USB device may consist of several logical sub-devices that are referred to as *device functions*. A single device may provide several functions, for example, a webcam (video device function) with a built-in microphone (audio device function). Such a device is called a *compound device* in which each logical device is assigned a distinctive address by the host and all logical devices are connected to a built-in hub to which the physical USB wire is connected. A host assigns one and only one device address to a function.

USB device communication is based on *pipes* (logical channels). A pipe is a connection from the host controller to a logical entity, found on a device, and named an *endpoint*. Because pipes correspond 1-to-1 to endpoints, the terms are sometimes used interchangeably. A USB device can have up to 32 endpoints: 16 into the host controller and 16 out of the host controller. The USB standard reserves one endpoint of each type, leaving a theoretical maximum of 30 for normal use. USB devices seldom have this many endpoints.

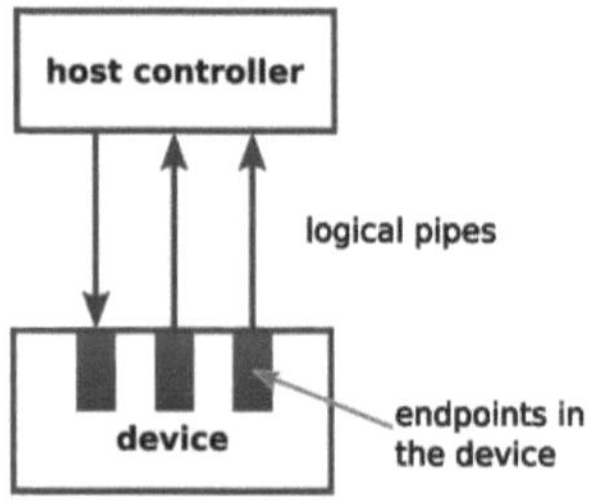

USB endpoints actually reside on the connected device: the channels to the host are referred to as pipes

There are two types of pipes: stream and message pipes depending on the type of data transfer.

- *isochronous transfers*: at some guaranteed data rate (often, but not necessarily, as fast as possible) but with possible data loss (e.g., realtime audio or video).
- *interrupt transfers*: devices that need guaranteed quick responses (bounded latency) (e.g., pointing devices and keyboards).
- *bulk transfers*: large sporadic transfers using all remaining available bandwidth, but with no guarantees on bandwidth or latency (e.g., file transfers).
- *control transfers*: typically used for short, simple commands to the device, and a status response, used, for example, by the bus control pipe number 0.

A stream pipe is a uni-directional pipe connected to a uni-directional endpoint that transfers data using an *isochronous*, *interrupt*, or *bulk* transfer. A message pipe is a bi-directional pipe connected to a bi-directional endpoint that is exclusively used for *control* data flow. An endpoint is built into the USB device by the manufacturer and therefore exists permanently. An endpoint of a pipe is addressable with a tuple *(device_address, endpoint_number)* as specified in a TOKEN packet that the host sends when it wants to start a data transfer session. If the direction of the data transfer is from the host to the endpoint, an OUT packet (a specialization of a TOKEN packet) having the desired device address and endpoint number is sent by the host. If the direction of the data transfer is from the device to the host, the host sends an IN packet instead. If the destination endpoint is a uni-directional endpoint whose manufacturer's designated direction does not match the TOKEN packet (e.g., the manufacturer's designated direction is IN while the TOKEN packet is an OUT packet), the TOKEN packet will be ignored. Otherwise, it will be accepted and the data transaction can start. A bi-directional endpoint, on the other hand, accepts both IN and OUT packets.

Two USB receptacles on the front of a computer

Endpoints are grouped into *interfaces* and each interface is associated with a single device function. An exception to this is endpoint zero, which is used for device configuration and which is not associated with any interface. A single device function composed of independently controlled interfaces is called a *composite device*. A composite device only has a single device address because the host only assigns a device address to a function.

When a USB device is first connected to a USB host, the USB device enumeration process is started. The enumeration starts by sending a reset signal to the USB device. The data rate of the USB device is determined during the reset signaling. After reset, the USB device's information is read by the host and the device is assigned a unique 7-bit address. If the device is supported by the host, the device drivers needed for communicating with the device are loaded and the device is set to a configured state. If the USB host is restarted, the enumeration process is repeated for all connected devices.

The host controller directs traffic flow to devices, so no USB device can transfer any data on the bus without an explicit request from the host controller. In USB 2.0, the host controller polls the bus for traffic, usually in a round-robin fashion. The slowest device connected to a controller sets the bandwidth of the interface. For *SuperSpeed USB* (defined since USB 3.0), connected devices can request service from host. Because there are two separate controllers in each USB 3.0 host, USB 3.0 devices will transmit and receive at USB 3.0 data rates regardless of USB 2.0 or earlier devices connected to that host. Operating data rates for them will be set in the legacy manner.

Device classes

USB defines class codes used to identify a device's functionality and to load a device driver based on that functionality. This enables every device driver writer to support devices from different manufacturers that comply with a given class code.

Device classes include:[13]

Class	Usage	Description	Examples, or exception
00h	Device	Unspecified[14]	Device class is unspecified, interface descriptors are used to determine needed drivers
01h	Interface	Audio	Speaker, microphone, sound card, MIDI
02h	Both	Communications and CDC Control	Modem, Ethernet adapter, Wi-Fi adapter
03h	Interface	Human interface device (HID)	Keyboard, mouse, joystick
05h	Interface	Physical Interface Device (PID)	Force feedback joystick
06h	Interface	Image	Webcam, scanner
07h	Interface	Printer	Laser printer, inkjet printer, CNC machine
08h	Interface	Mass storage	USB flash drive, memory card reader, digital audio player, digital camera, external drive
09h	Device	USB hub	Full bandwidth hub
0Ah	Interface	CDC-Data	Used together with class 02h: communications and CDC control
0Bh	Interface	Smart Card	USB smart card reader
0Dh	Interface	Content security	Fingerprint reader
0Eh	Interface	Video	Webcam
0Fh	Interface	Personal Healthcare	Pulse monitor (watch)
DCh	Both	Diagnostic Device	USB compliance testing device
E0h	Interface	Wireless Controller	Bluetooth adapter, Microsoft RNDIS
EFh	Both	Miscellaneous	ActiveSync device
FEh	Interface	Application-specific	IrDA Bridge, Test & Measurement Class (USBTMC),[15] USB DFU (Direct Firmware update)[16]
FFh	Both	Vendor-specific	Indicates that a device needs vendor specific drivers

USB mass storage

A flash drive, a typical USB mass-storage device

USB implements connections to storage devices using a set of standards called the *USB mass storage device class* (referred to as MSC or UMS). This was initially intended for traditional magnetic and optical drives, but has been extended to support a wide variety of devices, particularly flash drives. This generality is because many systems can be controlled with the familiar metaphor of file manipulation within directories (the process of making a novel device look like a familiar device is also known as extension). The ability to boot a write-locked SD card with a USB adapter is particularly advantageous for maintaining the integrity and non-corruptible, pristine state of the booting medium.

Though most newer computers are capable of booting off USB mass storage devices, USB is not intended to be a primary bus for a computer's internal storage: buses such as Parallel ATA (PATA or IDE), Serial ATA (SATA), or SCSI fulfill that role in PC class computers. However, USB has one important advantage in that it is possible to install and remove devices without rebooting the computer (hot-swapping), making it useful for mobile peripherals, including drives of various kinds. Originally conceived and still used today for optical storage devices (CD-RW drives, DVD drives and so on), several manufacturers offer external portable USB hard disk drives, or empty enclosures for disk drives, which offer performance comparable to internal drives, limited by the current number and type of attached USB devices and by the upper limit of the USB interface (in practice about 30 MB/s for USB 2.0 and potentially 400 MB/s or more[17] for USB 3.0). These external drives have typically included a "translating device" that bridges between a drive's interface to a USB interface port. Functionally, the drive appears to the user much like an internal drive. Other competing standards for external drive connectivity include eSATA, ExpressCard (now at version 2.0), and FireWire (IEEE 1394).

Another use for USB mass storage devices is the portable execution of software applications (such as web browsers and VoIP clients) with no need to install them on the host computer.[18] [19]

Human interface devices (HIDs)

Joysticks, keypads, tablets and other human-interface devices are also progressively migrating from MIDI, and PC game port connectors to USB.

USB mice and keyboards can usually be used with older computers that have PS/2 connectors with the aid of a small USB-to-PS/2 adapter. Such adaptors contain no logic circuitry: the hardware in the USB keyboard or mouse is designed to detect whether it is connected to a USB or PS/2 port, and communicate using the appropriate protocol. Converters also exist to allow PS/2 keyboards and mice (usually one of each) to be connected to a USB port. These devices present two HID endpoints to the system and use a microcontroller to perform bidirectional translation of data between the two standards.

Physical appearance

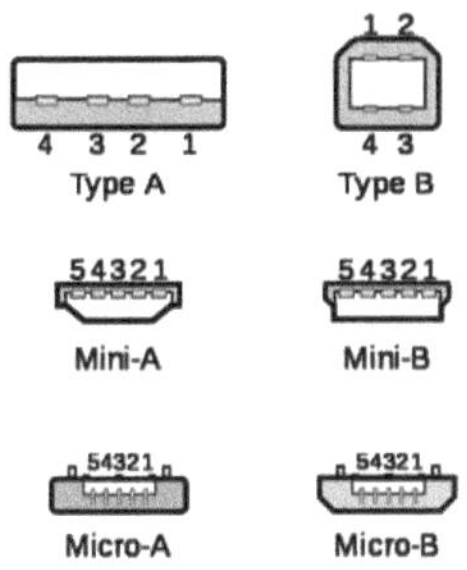

Pinouts of Standard, Mini, and Micro USB plugs. The USB logo is on the bottom of the two micro-USB plugs (as they are shown in this figure) but on the top of the other plugs[20]

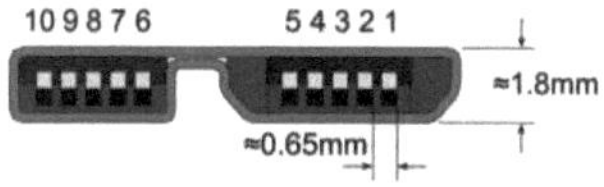

Micro-B USB 3.0 compatible socket

USB 2.0 connector on the side of the specification standard micro USB 3.0 connector are aligned pin-minute increase in the standard.

No.1: power (VBUS)

No.2: USB 2.0 differential pair (D–)

No.3: USB 2.0 differential pair (D+)

No.4: USB OTG ID for identifying lines

No.5: GND

No.6: USB 3.0 signal transmission line (–)

No.7: USB 3.0 signal transmission line (+)

No.8: GND

No.9: USB 3.0 signal receiving line (–)

No.10: USB 3.0 signal receiving line (+)

USB 1.x/2.0 standard pinout

Pin	Name	Cable color	Description
1	VBUS	Red	+5 V
2	D–	White	Data –
3	D+	Green	Data +
4	GND	Black	Ground

USB 1.x/2.0 Mini/Micro pinout

Pin	Name	Cable color	Description
1	VBUS	Red	+5 V
2	D–	White	Data –
3	D+	Green	Data +
4	ID	None	Permits distinction of A plug from B plug * A plug: connected to Signal ground * B plug: not connected
5	GND	Black	Signal ground

Connector properties

The connectors specified by the USB committee were designed to support a number of USB's underlying goals, and to reflect lessons learned from the menagerie of connectors which have been used in the computer industry. The connector mounted on the host or device is called the **receptacle**, and the connector attached to the cable is called the **plug**.[21] In the case of an extension cable, the connector on one end is a receptacle. The official USB specification documents periodically define the term **male** to represent the plug, and **female** to represent the receptacle.

Standard type A plug and receptacle

Usability and "upside down" connectors

By design, it is difficult to attach a USB connector incorrectly. Connectors cannot be plugged in upside down and it is clear from the physical act of making a connection, when the plug and receptacle are correctly mated. The USB specification states that the required USB Icon is to be "embossed" on the "topside" of the USB plug, which "provides easy user recognition and facilitates alignment during the mating process". The specification also shows that the "recommended" (optional) "Manufacturer's logo" ("engraved" on the diagram but not specified in the text) is on the opposite side of the USB Icon. The specification further states "the USB Icon is also located adjacent to each receptacle. Receptacles should be oriented to allow the icon on the plug to be visible during the mating process". However, the specification does not consider the height of the device compared to the eye level height of the user, so the side of the cable that is "visible" when mated to a computer on a desk can depend on whether the user is standing or kneeling.[21]

USB extension cord

Only moderate insertion/removal force is needed. USB cables and small USB devices are held in place by the gripping force from the receptacle (without need of the screws, clips, or thumb-turns other connectors have required). The force needed to make or break a connection is modest, allowing connections to be made in awkward circumstances (i.e., behind a floor-mounted chassis, or from below) or by those with motor disabilities.

The standard connectors were deliberately intended to enforce the directed topology of a USB network: type A connectors on host devices that supply power and type B connectors on target devices that receive power. This prevents users from accidentally connecting two USB power supplies to each other, which could lead to dangerously high currents, circuit failures, or even fire. USB does not support cyclical networks and the standard connectors from incompatible USB devices are themselves incompatible. Unlike other communications systems (e.g. network cabling) gender changers make little sense with USB and are almost never used.

Durability

The standard connectors were designed to be robust. Many previous connector designs were fragile, specifying embedded component pins or other delicate parts which proved vulnerable to bending or breakage, even with the application of modest force. The electrical contacts in a USB connector are protected by an adjacent plastic tongue, and the entire connecting assembly is usually protected by an enclosing metal sheath.

The connector construction always ensures that the external sheath on the plug makes contact with its counterpart in the receptacle before any of the four connectors within make electrical contact. The external metallic sheath is typically connected to system ground, thus dissipating damaging static charges. This enclosure design also provides a degree of protection from electromagnetic interference to the USB signal while it travels through the mated connector pair (the only location when the otherwise twisted data pair travels in parallel). In addition, because of the required sizes of the power and common connections, they are made after the system ground but before the data connections. This type of staged make-break timing allows for electrically safe hot-swapping.

The newer Micro-USB receptacles are designed for up to 10,000 cycles of insertion and removal between the receptacle and plug, compared to 1,500 for the standard USB and 5,000 for the Mini-USB receptacle. This is accomplished by adding a locking device and by moving the leaf-spring connector from the jack to the plug, so that the most-stressed part is on the cable side of the connection. This change was made so that the connector on the less expensive cable would bear the most wear instead of the more expensive micro-USB device.

Compatibility

The USB standard specifies relatively loose tolerances for compliant USB connectors to minimize physical incompatibilities in connectors from different vendors. To address a weakness present in some other connector standards, the USB specification also defines limits to the size of a connecting device in the area around its plug. This was done to prevent a device from blocking adjacent ports due to the size of the cable strain relief mechanism (usually molding integral with the cable outer insulation) at the connector. Compliant devices must either fit within the size restrictions or support a compliant extension cable which does. Two-way communication is also possible. In USB 3.0, full-duplex communications are done when using SuperSpeed (USB 3.0) transfer. In previous USB versions (i.e., 1.x or 2.0), all communication is half-duplex and directionally controlled by the host.

In general, cables have only plugs (very few have a receptacle on one end, although extension cables with a standard A plug and jack are sold), and hosts and devices have only receptacles. Hosts almost universally have type-A receptacles, and devices one or another type-B variety. Type-A plugs mate only with type-A receptacles, and type-B with type-B; they are deliberately physically incompatible. However, an extension to USB standard specification called USB On-The-Go allows a single port to act as either a host or a device—chosen by which end of the cable plugs into the receptacle on the unit. Even after the cable is hooked up and the units are communicating, the two units may "swap" ends under program control. This capability is meant for units such as PDAs in which the USB link might connect to a PC's host port as a device in one instance, yet connect as a host itself to a keyboard and mouse device in another instance.

USB 3.0 receptacles are electrically compatible with USB Standard 2.0 device plugs if they physically match. USB 3.0 type-A plugs and receptacles are completely backward compatible, and USB 3.0 type-B receptacles will accept USB 2.0 and earlier plugs. However, USB 3.0 type-B plugs will not fit into USB 2.0 and earlier receptacles. eSATAp (eSATA/USB) port is also compatible with USB 2.0 devices.

Connector types

There are several types of USB connectors, including some that have been added while the specification progressed. The original USB specification detailed Standard-A and Standard-B plugs and receptacles. The first engineering change notice to the USB 2.0 specification added Mini-B plugs and receptacles.

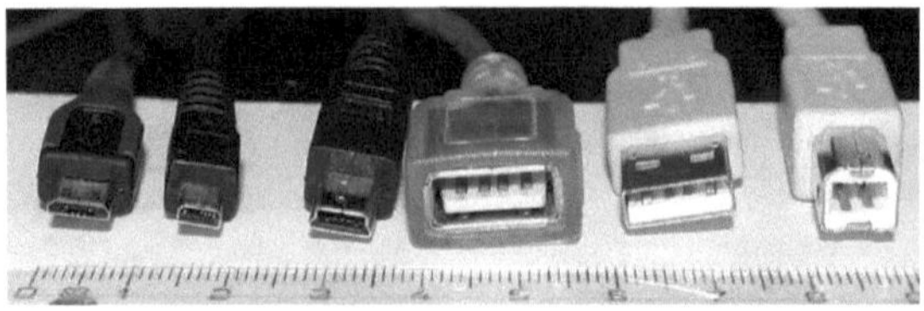

Types of USB connectors left to right (ruler in centimeters): Micro-B plug, ExtUSB plug (HTC proprietary variant), Mini-B plug (5-pin) Standard-A receptacle, Standard-A plug, Standard-B plug.

The data connectors in the Standard-A plug are actually recessed in the plug as compared to the outside power connectors. This permits the power to connect first which prevents data errors by allowing the device to power up first and then transfer the data. Some devices will operate in different modes depending on whether the data connection is made. This difference in connection can be exploited by inserting the connector only partially. For example, some battery-powered MP3 players switch into file transfer mode and cannot play MP3 files while a USB plug is fully inserted, but can be operated in MP3 playback mode using USB power by inserting the plug only part way so that the power slots make contact while the data slots do not. This enables those devices to be operated in MP3 playback mode while getting power from the cable.

To reliably enable a charge-only feature, modern USB accessory peripherals now include charging cables that provide power connections to the host port but no data connections, and both home and vehicle charging docks are available that supply power from a converter device and do not include a host device and data pins, allowing any capable USB device to be charged and/or operated from a standard USB cable.

USB standard connectors

The USB 2.0 Standard-A type of USB plug is a flattened rectangle which inserts into a "downstream-port" receptacle on the USB host, or a hub, and carries both power and data. This plug is frequently seen on cables that are permanently attached to a device, such as one connecting a keyboard or mouse to the computer via usb connection.

USB connections eventually wear out as the connection loosens through repeated plugging and unplugging. The lifetime of a USB-A male connector is approximately 1,500 connect/disconnect cycles.[22]

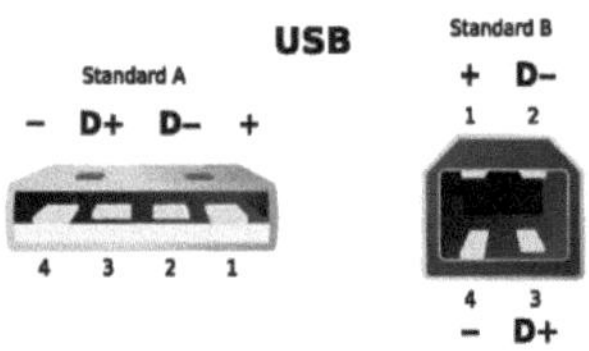

Pin configuration of the USB connectors Standard A/B, viewed looking into face/end of plug

A Standard-B plug—which has a square shape with bevelled exterior corners—typically plugs into an "upstream receptacle" on a device that uses a removable cable, e.g. a printer. A Type B plug delivers power in addition to carrying data. On some devices, the Type B receptacle has no data connections, being used solely for accepting power from the upstream device. This two-connector-type scheme (A/B) prevents a user from accidentally creating an electrical loop.[23]

Mini and Micro connectors

Micro B USB

Various connectors have been used for smaller devices such as PDAs, mobile phones or digital cameras. These include the now-deprecated[24] (but standardized) Mini-A and the currently standard Mini-B,[25] Micro-A, and Micro-B connectors.

The Mini-A and Mini-B plugs are approximately 3 by 7 mm. The micro-USB plugs have a similar width and approximately half the thickness, enabling their integration into thinner portable devices. The mini-A connector is 6.85 by 1.8 mm with a maximum overmold size of 11.7 by 8.5 mm. The mini-B connector is 6.85 by 1.8 mm with a maximum overmold size of 10.6 by 8.5 mm.[20]

USB Mini A (left) and USB Mini B (right) plugs

The Micro-USB connector was announced by the USB-IF on 4 January 2007.[26] The Mini-A connector and the Mini-AB receptacle connector were deprecated on 23 May 2007.[27] As of February 2009, many currently available devices and cables still use Mini plugs, but the newer Micro connectors are being widely adopted and as of December 2010, the Micro connectors are the most widely used. The thinner micro connectors are intended to replace the Mini plugs in new devices including smartphones and personal digital assistants. The Micro plug design is rated for at least 10,000 connect-disconnect cycles which is significantly more than the Mini plug design.[28] The *Universal Serial Bus Micro-USB Cables and Connectors Specification* details the mechanical characteristics of Micro-A plugs, Micro-AB receptacles, and Micro-B plugs and receptacles,[28] along with a Standard-A receptacle to Micro-A plug adapter.

The cellular phone carrier group, Open Mobile Terminal Platform (OMTP) in 2007 have endorsed Micro-USB as the standard connector for data and power on mobile devices.[29] These include various types of battery chargers, allowing Micro-USB to be the single external cable link needed by some devices.

As of 30 January 2009 Micro-USB has been accepted and is being used by almost all cell phone manufacturers as the standard charging port (including HTC, Motorola, Nokia, LG, Hewlett-Packard, Samsung, Sony Ericsson, Research In Motion) in most of the world.

On 29 June 2009, following a request from the European Commission and in close co-operation with the Commission services, major producers of mobile phones have agreed in a Memorandum of Understanding ("MoU") to harmonise chargers for data-enabled mobile phones sold in the European Union. Industry commits to provide charger compatibility on the basis of the Micro-USB connector. Consumers will be able to purchase mobile phones without a charger, thus logically reducing their cost.[30] Following a mandate from the European Commission, the European Standardisation Bodies CEN-CENELEC and ETSI have now made available the harmonised standards needed for the manufacture of data-enabled mobile phones compatible with the new common External Power Supply (EPS) based on micro-USB.[31]

In addition, on 22 October 2009 the International Telecommunication Union (ITU) has also announced that it had embraced micro-USB as the *Universal Charger Solution* its "energy-efficient one-charger-fits-all new mobile phone solution", and added: "Based on the Micro-USB interface, UCS chargers will also include a 4-star or higher efficiency rating—up to three times more energy-efficient than an unrated charger".[32]

A USB On-The-Go device is required to have one, and only one USB connector: a Mini-AB or Micro-AB receptacle. This receptacle is capable of accepting both Mini-A and Mini-B plugs, and alternatively, Micro-A and Micro-B plugs, attached to any of the legal cables and adapters as defined in Micro-USB1.01.

The OTG device with the A-plug inserted is called the A-device and is responsible for powering the USB interface when required and by default assumes the role of host. The OTG device with the B-plug inserted is called the B-device and by default assumes the role of peripheral. An OTG device with no plug inserted defaults to acting as a B-device. If an application on the B-device requires the role of host, then the HNP protocol is used to temporarily transfer the host role to the B-device.

OTG devices attached either to a peripheral-only B-device or a standard/embedded host will have their role fixed by the cable since in these scenarios it is only possible to attach the cable one way around.

Host interface receptacles

The following receptacles accept the following plugs:

Receptacle	Plug: Type A	Plug: Type B	Plug: Mini-B	Plug: Micro-A	Plug: Micro-B
Type A	Yes	No	No	No	No
Type B	No	Yes	No	No	No
Mini-B	No	No	Yes	No	No
Micro-AB	No	No	No	Yes	Yes
Micro-B	No	No	No	No	Yes

Cable plugs (USB 1.x/2.0)

Cables exist with pairs of plugs:

Plug	Plug: Micro-B	Plug: Micro-A	Plug: Mini-B	Plug: Type B	Plug: Type A
Type A	Yes	NS	Yes	Yes	NS
Type B	No	NS	No	No	
Mini-B	No	NS	No		
Micro-A	Yes	No			
Micro-B	No				

NS: non-standard, existing for specific proprietary purposes, and not interoperable with USB-IF compliant equipment.

In addition to the above cable assemblies comprising two plugs, an "adapter" cable with a Micro-A plug and a Standard-A receptacle is compliant with USB specifications.[20] Other combinations of connectors are not compliant.

However, some older devices and cables with Mini-A connectors have been certified by USB-IF. The Mini-A connector has been deprecated: there will be no new certification of assemblies using Mini-A connector.[24]

Proprietary connectors and formats

HTC ExtUSB

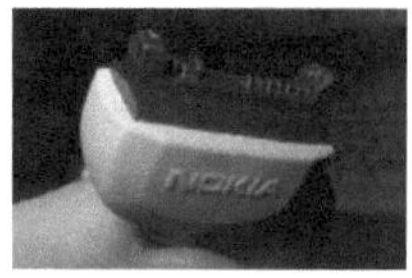
Nokia's discontinued Pop-Port connector

- Microsoft's original Xbox game console uses standard USB 1.1 signalling in its controllers and memory cards, but uses proprietary connectors and ports. The Xbox 360 (pre Xbox 360 S) has two Memory Unit ports which use USB signalling, but proprietary connectors and 3.3v power.[33]
- IBM UltraPort uses standard USB signalling, but via a proprietary connection format.
- American Power Conversion uses USB signalling and HID device class on its uninterruptible power supplies using 10P10C connectors.
- HTC manufactured Windows Mobile and Android-based devices which have a proprietary connector called HTC ExtUSB (Extended USB). ExtUSB combines mini-USB (with which it is backwards-compatible) with audio input as well as audio and video output in an 11-pin connector.
- Nokia included a USB connection as part of the Pop-Port connector on some older mobile phone models.
- Sony Ericsson used a proprietary connector called FastPort from 2005 to 2009.
- iriver added a fifth power pin within USB-A plugs for higher power and faster charging, used for the iriver U10 series. A mini-USB version contains a matching extra power pin for the cradle.
- Apple has shipped non-standard USB extension cables with some of their computers, for use with the included Apple USB keyboards. The extension cable's socket is keyed with a small protrusion to prevent the insertion of a standard USB plug, while the Apple USB keyboard's plug has a matching indentation. The indentation on the keyboard's plug does not interfere with insertion into a standard USB socket. Apple's proprietary 30-pin dock connector on its iPod, iPhone, and iPad serves purposes in addition to USB (such as analog audio, IEEE-1394, eSATA and HDMI). The second, third, and fourth generation iPod Shuffle uses a TRRS connector to carry USB, audio, or power signals.
- HP Tablet computers use non-standard connectors to transmit the USB signals between the keyboard/mouse unit and the Computer Tablet Unit.
- Some digital cameras have their own USB connectors, like the Panasonic Lumix DMC-FT2.
- The United States Army's Land Warrior system uses standard USB signaling with 15.6 V power using a ruggedized connector from Glenair, Inc.

Cabling

A USB twisted pair, where the "Data +" and "Data -" conductors are twisted together in a double helix. The wires are enclosed in a further layer of shielding.

The data cables for USB 1.x and USB 2.x use a twisted pair to reduce noise and crosstalk. USB 3.0 cables contain twice as many wires as USB 2.x to support SuperSpeed data transmission, and are thus larger in diameter.[34]

The USB 1.1 Standard specifies that a standard cable can have a maximum length of 3 meters with devices operating at Low Speed (1.5 Mb/s), and a maximum length of 5 meters with devices operating at Full Speed (12 Mb/s).

USB 2.0 provides for a maximum cable length of 5 meters for devices running at Hi Speed (480 Mb/s). The primary reason for this limit is the maximum allowed round-trip delay of about 1.5 μs. If USB host commands are unanswered by the USB device within the allowed time, the host considers the command lost. When adding USB device response time, delays from the maximum number of hubs added to the delays from connecting cables, the maximum acceptable delay per cable amounts to 26 ns.[35] The USB 2.0 specification requires cable delay to be less than 5.2 ns per meter (192,000 km/s, which is close to the maximum achievable transmission speed for standard copper wire).[36]

The USB 3.0 standard does not directly specify a maximum cable length, requiring only that all cables meet an electrical specification: for copper cabling with AWG 26 wires the maximum practical length is 3 metres (9.8 ft).[37]

Power

The USB 1.x and 2.0 specifications provide a 5 V supply on a single wire from which connected USB devices may draw power. The specification provides for no more than 5.25 V and no less than 4.75 V (5 V±5%) between the positive and negative bus power lines. For USB 3.0, the voltage supplied by low-powered hub ports is 4.45–5.25 V.[38]

A unit load is defined as 100 mA in USB 2.0, and 150 mA in USB 3.0. A device may draw a maximum of 5 unit loads (500 mA) from a port in USB 2.0; 6 (900 mA) in USB 3.0. There are two types of devices: low-power and high-power. A low-power device draws at most 1 unit load, with minimum operating voltage of 4.4 V in USB 2.0, and 4 V in USB 3.0. A high-power device draws the maximum number of unit loads permitted by the standard. Every device functions initially as low-power but the device may request high-power and will get it if the power is available on the providing bus.[39]

Some devices, such as high-speed external disk drives, require more than 500 mA of current[40] and therefore cannot be powered from one USB 2.0 port. Such devices usually come with Y-shaped cable that has two USB connectors to be plugged into a computer. With such a cable, a device can draw power from two USB ports simultaneously.[41]

A bus-powered hub initializes itself at 1 unit load and transitions to maximum unit loads after it completes hub configuration. Any device connected to the hub will draw 1 unit load regardless of the current draw of devices connected to other ports of the hub (i.e. one device connected on a four-port hub will draw only 1 unit load despite the fact that more unit loads are being supplied to the hub).[39]

A self-powered hub will supply maximum supported unit loads to any device connected to it. In addition, the V_{BUS} will present 1 unit load upstream for communication if parts of the Hub are powered down.[39]

Charging ports

In the *Battery Charging Specification*,[42] defined in 2007, new types of USB ports, *charging ports*, are defined. As compared to *standard downstream ports*, where a portable device can only draw more than 100 mA current after digital negotiation with the host or hub, charging ports can supply currents above 0.5 A without digital negotiation. A charging port will supply up to 500 mA at 5 V, up to the rated current at 3.6 V or more, and drop its output voltage if the portable device attempts to draw more than the rated current. The charger port may shut down if the load is too high.

Charging ports exist in two flavors: *downstream charging ports*, supporting data transfers as well, and *dedicated charging ports*, without data support. A portable device can recognize the type of USB port from the way the D+ and D- pins are connected. For example, on a dedicated charging port, the D+ and D- pins are shorted. With downstream charging ports, current passing through the thin ground wire may interfere with high-speed data signals. Therefore, current draw may not exceed 900 mA during high-speed data transfer. A dedicated charge port may have a rated current between 0.5 and 1.5 A. There is no upper limit for the rated current of a downstream charge port, as long as the connector can handle the current (standard USB 2.0 A-connectors are rated at 1.5 A).

Before the battery charging specification was defined, there was no standardized way for the portable device to inquire how much current was available. For example, Apple's iPod and iPhone chargers indicate the available current by voltages on the D- and D+ lines.[43]

Dedicated charging ports can be found on USB power adapters that convert utility power and/or another power source (e.g., a car's electrical system) to run attached devices and battery packs. On a host (such as a laptop computer) with both standard and charging USB ports, the charging ports should be labeled as such.[42]

Sleep-and-charge ports

Sleep-and-charge USB ports can be used to charge electronic devices even when the computer is switched off. Normally when a computer is powered off the USB ports are powered down. This prevents phones and other devices from being able to charge unless the computer is powered on. Sleep-and-charge USB ports remain powered even when the computer is off. On laptops, charging devices from the USB port when it is not being powered from AC will drain the laptop battery faster. Desktop machines need to remain plugged into AC power for Sleep-and-charge to work.[44]

Mobile device charger standards

As of 14 June 2007, all new mobile phones applying for a license in China are required to use the USB port as a power port.[45] [46] This was the first standard to use the convention of shorting D+ and D-.[47]

The Micro-USB interface is commonly found on chargers for mobile phones

In September 2007, the Open Mobile Terminal Platform group (a forum of mobile network operators and manufacturers such as Nokia, Samsung, Motorola, Sony Ericsson and LG) announced that its members had agreed on micro-USB as the future common connector for mobile devices.[48] [49]

On 17 February 2009, the GSM Association (GSMA) announced[50] that they had agreed on a standard charger for mobile phones. The standard connector to be adopted by 17 manufacturers including Nokia, Motorola and Samsung is to be the micro-USB connector (several media reports erroneously reported this as the mini-USB). The new chargers will be much more efficient than existing chargers.[50] Having a standard charger for all phones means that manufacturers will no longer have to supply a charger with every new phone. The basis of the GSMA's Universal Charger Solution (UCS) is the technical recommendation from OMTP and the USB-IF battery charging standard.[51] [52] [53]

On 22 April 2009, this was further endorsed by the CTIA – The Wireless Association.[54]

In June 2009, many of the world's largest mobile phone manufacturers signed a Memorandum of Understanding (MoU), agreeing to make most data-enabled mobile phones marketed in the European Union compatible with a common External Power Supply (EPS) based on the GSMA / OMTP Universal Charging Solution.[55] [56]

On 22 October 2009, the International Telecommunication Union (ITU) announced that it had embraced the Universal Charger Solution as its "energy-efficient one-charger-fits-all new mobile phone solution", and added: "Based on the Micro-USB interface, UCS chargers will also include a 4-star or higher efficiency rating—up to three times more energy-efficient than an unrated charger".[57]

Non-standard devices

Some USB devices require more power than is permitted by the specifications for a single port. This is common for external hard and optical disc drives, and generally for devices with motors or lamps. Such devices can use an external power supply, which is allowed by the standard, or use a dual-input USB cable, one input of which is used for power and data transfer, the other solely for power, which makes the device a non-standard USB device. Some USB ports and external hubs can, in practice, supply more power to USB devices than required by the specification but a standard-compliant device may not depend on this.

USB vacuum cleaner novelty device

In addition to limiting the total average power used by the device, the USB specification limits the inrush current (i.e., that used to charge decoupling and filter capacitors) when the device is first connected. Otherwise, connecting a device could cause problems with the host's internal power. USB devices are also required to automatically enter ultra low-power suspend mode when the USB host is suspended. Nevertheless, many USB host interfaces do not cut off the power supply to USB devices when they are suspended since resuming from the suspended state would become a lot more complicated if they did.

Some non-standard USB devices use the 5 V power supply without participating in a proper USB network which negotiates power draws with the host interface. These are usually referred to as USB decorations. The typical example is a USB-powered keyboard light; fans, mug coolers and heaters, battery chargers, miniature vacuum cleaners, and even miniature lava lamps are available. In most cases, these items contain no digital circuitry, and thus are not Standard compliant USB devices at all. This can theoretically cause problems with some computers, such as drawing too much current and damaging circuitry; prior to the Battery Charging Specification, the USB specification required that devices connect in a low-power mode (100 mA maximum) and communicate their current requirements to the host, which would then permit the device to switch into high-power mode.

Some devices, when plugged into charging ports, draw even more power (10 watts or 2.1 Amps) than the Battery Charging Specification allows. The iPad and MiFi 2200 are two such devices. [58]

Powered USB

Powered USB uses standard USB signaling with the addition of extra power lines. It uses four additional pins to supply up to 6 A at either 5 V, 12 V, or 24 V (depending on keying) to peripheral devices. The wires and contacts on the USB portion have been upgraded to support higher current on the 5 V line, as well. This is commonly used in retail systems and provides enough power to operate stationary barcode scanners, printers, PIN pads, signature capture devices, etc. This modification of the USB interface is proprietary and was developed by IBM, NCR, and FCI/Berg. It is essentially two connectors stacked such that the bottom connector accepts a standard USB plug and the top connector takes a power connector.

Signaling

USB supports the following signaling rates. The terms *speed* and *bandwidth* are used interchangeably. "high-" is alternatively written as "hi-".

- A *low-speed* rate of 1.5 Mbit/s (~183 kB/s) is defined by USB 1.0. It is very similar to full-bandwidth operation except each bit takes 8 times as long to transmit. It is intended primarily to save cost in low-bandwidth human interface devices (HID) such as keyboards, mice, and joysticks.
- The *full-speed* rate of 12 Mbit/s (~1.43 MB/s) is the basic USB data rate defined by USB 1.0. All USB hubs support full-bandwidth.

- A *high-speed* (USB 2.0) rate of 480 Mbit/s (~57 MB/s) was introduced in 2001. All hi-speed devices are capable of falling back to full-bandwidth operation if necessary; i.e., they are backward compatible with USB 1.1. Connectors are identical for USB 2.0 and USB 1.x.
- A *SuperSpeed* (USB 3.0) rate of 5.0 Gbit/s (~596 MB/s). The written USB 3.0 specification was released by Intel and partners in August 2008. The first USB 3 controller chips were sampled by NEC May 2009[59] and products using the 3.0 specification arrived beginning in January 2010.[60] USB 3.0 connectors are generally backwards compatible, but include new wiring and full duplex operation.

USB signals are transmitted on a twisted-pair data cable with 90Ω ±15% Characteristic impedance,[61] labeled D+ and D–. Prior to USB 3.0, these collectively use half-duplex differential signaling to reduce the effects of electromagnetic noise on longer lines. Transmitted signal levels are 0.0 to 0.3 volts for low and 2.8 to 3.6 volts for high in full-bandwidth and low-bandwidth modes, and −10 to 10 mV for low and 360 to 440 mV for high in hi-bandwidth mode. In FS mode, the cable wires are not terminated, but the HS mode has termination of 45 Ω to ground, or 90 Ω differential to match the data cable impedance, reducing interference due to signal reflections. USB 3.0 introduces two additional pairs of shielded twisted wire and new, mostly interoperable contacts in USB 3.0 cables, for them. They permit the higher data rate, and full duplex operation.

A USB connection is always between a host or hub at the "A" connector end, and a device or hub's "upstream" port at the other end. Originally, this was a "B' connector, preventing erroneous loop connections, but additional upstream connectors were specified, and some cable vendors designed and sold cables which permitted erroneous connections (and potential damage to the circuitry). USB interconnections are not as fool-proof or as simple as originally intended.

The host includes 15 kΩ pull-down resistors on each data line. When no device is connected, this pulls both data lines low into the so-called "single-ended zero" state (SE0 in the USB documentation), and indicates a reset or disconnected connection.

A USB device pulls one of the data lines high with a 1.5 kΩ resistor. This overpowers one of the pull-down resistors in the host and leaves the data lines in an idle state called "J". For USB 1.x, the choice of data line indicates a device's bandwidth support; full-bandwidth devices pull D+ high, while low-bandwidth devices pull D– high.

USB data is transmitted by toggling the data lines between the J state and the opposite K state. USB encodes data using the NRZI convention; a 0 bit is transmitted by toggling the data lines from J to K or vice-versa, while a 1 bit is transmitted by leaving the data lines as-is. To ensure a minimum density of signal transitions remains in the bitstream, USB uses bit stuffing; an extra 0 bit is inserted into the data stream after any appearance of six consecutive 1 bits. Seven consecutive received 1 bits is always an error. USB 3.0 has introduced additional data transmission encodings.

A USB packet begins with an 8-bit synchronization sequence '00000001'. That is, after the initial idle state J, the data lines toggle KJKJKJKK. The final 1 bit (repeated K state) marks the end of the sync pattern and the beginning of the USB frame. For high bandwidth USB, the packet begins with a 32-bit synchronization sequence.

A USB packet's end, called EOP (end-of-packet), is indicated by the transmitter driving 2 bit times of SE0 (D+ and D– both below max) and 1 bit time of J state. After this, the transmitter ceases to drive the D+/D– lines and the aforementioned pull up resistors hold it in the J (idle) state. Sometimes skew due to hubs can add as much as one bit time before the SE0 of the end of packet. This extra bit can also result in a "bit stuff violation" if the six bits before it in the CRC are '1's. This bit should be ignored by receiver.

A USB bus is reset using a prolonged (10 to 20 milliseconds) SE0 signal.

USB 2.0 devices use a special protocol during reset, called "chirping", to negotiate the high bandwidth mode with the host/hub. A device that is HS capable first connects as an FS device (D+ pulled high), but upon receiving a USB RESET (both D+ and D– driven LOW by host for 10 to 20 ms) it pulls the D– line high, known as chirp K. This indicates to the host that the device is high bandwidth. If the host/hub is also HS capable, it chirps (returns alternating J and K states on D– and D+ lines) letting the device know that the hub will operate at high bandwidth. The device has to receive at least 3 sets of KJ chirps before it changes to high bandwidth terminations and begins high bandwidth signaling. Because USB 3.0 uses wiring separate and additional to that used by USB 2.0 and USB 1.x, such bandwidth negotiation is not required.

Clock tolerance is 480.00 Mbit/s ±500 ppm, 12.000 Mbit/s ±2500 ppm, 1.50 Mbit/s ±15000 ppm.

Though high bandwidth devices are commonly referred to as "USB 2.0" and advertised as "up to 480 Mbit/s", not all USB 2.0 devices are high bandwidth. The USB-IF certifies devices and provides licenses to use special marketing logos for either "basic bandwidth" (low and full) or high bandwidth after passing a compliance test and paying a licensing fee. All devices are tested according to the latest specification, so recently compliant low bandwidth devices are also 2.0 devices.

USB 3 uses tinned copper stranded AWG-28 cables with 90 ± 7 Ω impedance for its high-speed differential pairs and linear feedback shift register and 8b/10b encoding sent with a voltage of 1 V nominal with a 100 mV receiver threshold; the receiver uses equalization.[62] SSC clock and 300 ppm precision is used. Packet headers are protected with CRC-16, while data payload is protected with CRC-32.[63] Power up to 3.6 W may be used. One unit load in superspeed mode is equal to 150 mA.[63]

Transfer rates

The theoretical maximum data rate in USB 2.0 is 480 Mbit/s (60 MB/s) per controller and is shared amongst all attached devices. Some chipset manufacturers overcome this bottleneck by providing multiple USB 2.0 controllers within the southbridge.

Typical hi-speed USB hard drives can be written to at rates around 25–30 MB/s, and read from at rates of 30–42 MB/s, according to routine testing done by CNet.[64] This is 70% of the total bandwidth available.

According to a USB-IF chairman, "at least 10 to 15 percent of the stated peak 60 MB/s (480 Mbit/s) of Hi-Speed USB goes to overhead—the communication protocol between the card and the peripheral. Overhead is a component of all connectivity standards".[65] Tables illustrating the transfer limits are shown in Chapter 5 of the USB spec.

For isochronous devices like audio streams, the bandwidth is constant, and reserved exclusively for a given device. The bus bandwidth therefore only has an effect on the number of channels that can be sent at a time, not the "speed" or latency of the transmission.

Communication

USB communication takes the form of packets. Initially, all packets are sent from the host, via the root hub and possibly more hubs, to devices. Some of those packets direct a device to send some packets in reply.

After the sync field, all packets are made of 8-bit bytes, transmitted least-significant bit first. The first byte is a packet identifier (PID) byte. The PID is actually 4 bits; the byte consists of the 4-bit PID followed by its bitwise complement. This redundancy helps detect errors. (Note also that a PID byte contains at most four consecutive 1 bits, and thus will never need bit-stuffing, even when combined with the final 1 bit in the sync byte. However, trailing 1 bits in the PID may require bit-stuffing within the first few bits of the payload.)

USB PID bytes

Type	PID value (msb-first)	Transmitted byte (lsb-first)	Name	Description
Reserved	0000	0000 1111		
Token	1000	0001 1110	**SPLIT**	High-bandwidth (USB 2.0) split transaction
	0100	0010 1101	**PING**	Check if endpoint can accept data (USB 2.0)
Special	1100	0011 1100	**PRE**	Low-bandwidth USB preamble
Handshake			**ERR**	Split transaction error (USB 2.0)
	0010	0100 1011	**ACK**	Data packet accepted
	1010	0101 1010	**NAK**	Data packet not accepted; please retransmit
	0110	0110 1001	**NYET**	Data not ready yet (USB 2.0)
	1110	0111 1000	**STALL**	Transfer impossible; do error recovery
Token	0001	1000 0111	**OUT**	Address for host-to-device transfer
	1001	1001 0110	**IN**	Address for device-to-host transfer
	0101	1010 0101	**SOF**	Start of frame marker (sent each ms)
	1101	1011 0100	**SETUP**	Address for host-to-device control transfer
Data	0011	1100 0011	**DATA0**	Even-numbered data packet
	1011	1101 0010	**DATA1**	Odd-numbered data packet
	0111	1110 0001	**DATA2**	Data packet for high-bandwidth isochronous transfer (USB 2.0)
	1111	1111 0000	**MDATA**	Data packet for high-bandwidth isochronous transfer (USB 2.0)

Packets come in three basic types, each with a different format and CRC (cyclic redundancy check):

Handshake packets

Handshake packets consist of nothing but a PID byte, and are generally sent in response to data packets. The three basic types are *ACK*, indicating that data was successfully received, *NAK*, indicating that the data cannot be received at this time and should be retried, and *STALL*, indicating that the device has an error and will never be able to successfully transfer data until some corrective action (such as device initialization) is performed.

USB 2.0 added two additional handshake packets, *NYET* which indicates that a split transaction is not yet complete. A NYET packet is also used to tell the host that the receiver has accepted a data packet, but cannot accept any more due to buffers being full. The host will then send PING packets and will continue with data packets once the device ACK's the PING. The other packet added was the *ERR* handshake to indicate that a split transaction failed.

The only handshake packet the USB host may generate is ACK; if it is not ready to receive data, it should not instruct a device to send any.

Token packets

Token packets consist of a PID byte followed by 2 payload bytes: 11 bits of address and a 5-bit CRC. Tokens are only sent by the host, never a device.

IN and *OUT* tokens contain a 7-bit device number and 4-bit function number (for multifunction devices) and command the device to transmit DATAx packets, or receive the following DATAx packets, respectively.

An IN token expects a response from a device. The response may be a NAK or STALL response, or a DATAx frame. In the latter case, the host issues an ACK handshake if appropriate.

An OUT token is followed immediately by a DATAx frame. The device responds with ACK, NAK, NYET, or STALL, as appropriate.

SETUP operates much like an OUT token, but is used for initial device setup. It is followed by an 8-byte DATA0 frame with a standardized format.

Every millisecond (12000 full-bandwidth bit times), the USB host transmits a special *SOF* (start of frame) token, containing an 11-bit incrementing frame number in place of a device address. This is used to synchronize isochronous data flows. High-bandwidth USB 2.0 devices receive 7 additional duplicate SOF tokens per frame, each introducing a 125 µs "microframe" (60000 high-bandwidth bit times each).

USB 2.0 added a *PING* token, which asks a device if it is ready to receive an OUT/DATA packet pair. The device responds with ACK, NAK, or STALL, as appropriate. This avoids the need to send the DATA packet if the device knows that it will just respond with NAK.

USB 2.0 also added a larger 3-byte *SPLIT* token with a 7-bit hub number, 12 bits of control flags, and a 5-bit CRC. This is used to perform split transactions. Rather than tie up the high-bandwidth USB bus sending data to a slower USB device, the nearest high-bandwidth capable hub receives a SPLIT token followed by one or two USB packets at high bandwidth, performs the data transfer at full or low bandwidth, and provides the response at high bandwidth when prompted by a second SPLIT token.

Data packets

A data packet consists of the PID followed by 0–1,023 bytes of data payload (up to 1,024 in high bandwidth, at most 8 at low bandwidth), and a 16-bit CRC.

There are two basic data packets, *DATA0* and *DATA1*. They must always be preceded by an address token, and are usually followed by a handshake token from the receiver back to the transmitter. The two packet types provide the 1-bit sequence number required by Stop-and-wait ARQ. If a USB host does not receive a response (such as an ACK) for data it has transmitted, it does not know if the data was received or not; the data might have been lost in transit, or it might have been received but the handshake response was lost.

To solve this problem, the device keeps track of the type of DATAx packet it last accepted. If it receives another DATAx packet of the same type, it is acknowledged but ignored as a duplicate. Only a DATAx packet of the opposite type is actually received.

When a device is reset with a SETUP packet, it expects an 8-byte DATA0 packet next.

USB 2.0 added *DATA2* and *MDATA* packet types as well. They are used only by high-bandwidth devices doing high-bandwidth isochronous transfers which need to transfer more than 1024 bytes per 125 µs microframe (8,192 kB/s).

PRE "packet"

Low-bandwidth devices are supported with a special PID value, *PRE*. This marks the beginning of a low-bandwidth packet, and is used by hubs which normally do not send full-bandwidth packets to low-bandwidth devices. Since all PID bytes include four 0 bits, they leave the bus in the full-bandwidth K state, which is the same as the low-bandwidth J state. It is followed by a brief pause during which hubs enable their low-bandwidth outputs, already idling in the J state, then a low-bandwidth packet follows, beginning with a sync sequence and PID byte, and ending with a brief period of SE0. Full-bandwidth devices other than hubs can simply ignore the PRE packet and its low-bandwidth contents, until the final SE0 indicates that a new packet follows.

Comparisons with other connection methods

FireWire

USB was originally seen as a complement to FireWire (IEEE 1394), which was designed as a high-bandwidth serial bus which could efficiently interconnect peripherals such as hard disks, audio interfaces, and video equipment. USB originally operated at a far lower data rate and used much simpler hardware, and was suitable for small peripherals such as keyboards and mice.

The most significant technical differences between FireWire and USB include the following:

- USB networks use a tiered-star topology, while FireWire networks use a tree topology.
- USB 1.0, 1.1 and 2.0 use a "speak-when-spoken-to" protocol; peripherals cannot communicate with the host unless the host specifically requests communication. USB 3.0 allows for device-initiated communications towards the host. A FireWire device can communicate with any other node at any time, subject to network conditions.
- A USB network relies on a single host at the top of the tree to control the network. In a FireWire network, any capable node can control the network.
- USB runs with a 5 V power line, while Firewire in current implementations supplies 12 V and theoretically can supply up to 30 V.
- Standard USB hub ports can provide from the typical 500 mA/2.5 W of current, only 100 mA from non-hub ports. USB 3.0 and USB On-The-Go supply 1.8 A/9.0 W (for dedicated battery charging, 1.5 A/7.5 W Full bandwidth or 900 mA/4.5 W High Bandwidth), while FireWire can in theory supply up to 60 watts of power, although 10 to 20 watts is more typical.

These and other differences reflect the differing design goals of the two buses: USB was designed for simplicity and low cost, while FireWire was designed for high performance, particularly in time-sensitive applications such as audio and video. Although similar in theoretical maximum transfer rate, FireWire 400 is faster than USB 2.0 Hi-Bandwidth in real-use,[66] especially in high-bandwidth use such as external hard-drives.[67] [68] [69] [70] The newer FireWire 800 standard is twice as fast as FireWire 400 and faster than USB 2.0 Hi-Bandwidth both theoretically and practically.[71] The chipset and drivers used to implement USB and Firewire have a crucial impact on how much of the bandwidth prescribed by the specification is achieved in the real world, along with compatibility with peripherals.[72]

Ethernet

The IEEE 802.3af Power over Ethernet (PoE) standard has a more elaborate power negotiation scheme than powered USB. It operates at 48 V DC and can supply more power (up to 12.95 W, PoE+ 25.5 W) over a cable up to 100 meters compared to USB 2.0 which provide 2.5 W with a maximum cable length of 5 meters. This has made PoE popular for VoIP telephones, security cameras, wireless access points and other networked devices within buildings. However, USB is cheaper than PoE provided that the distance is short, and power demand is low.

Ethernet standards requires electrical isolation between the networked device (computer, phone, etc.) and the network cable up to 1500 V AC or 2250 V DC for 60 seconds.[73] USB has no such requirement as it was designed for peripherals closely associated with a host computer, and in fact it connects the peripheral and host grounds. This gives Ethernet a significant safety advantage over USB with peripherals such as cable and DSL modems connected to external wiring that can assume hazardous voltages under certain fault conditions.[74]

Digital musical instruments

Digital musical instruments are another example of where USB is competitive for low-cost devices. However Power over Ethernet and the MIDI plug standard are preferred in high-end devices that must work with long cables. USB can cause ground loop problems in equipment because it connects the ground wires on both transceivers. By contrast, the MIDI plug standard and Ethernet have built-in isolation to 500 V or more.

eSATA/eSATAp

The eSATA connector is a more robust SATA connector, intended for connection to external hard drives and SSDs. It has a far higher transfer rate (3 Gbit/s or 6 Gbit/s, bi-directional) than USB 2.0. A device connected by eSATA appears as an ordinary SATA device, giving both full performance and full compatibility associated with internal drives.

eSATA does not supply power to external devices. This is an increasing disadvantage compared to USB. Even though USB's 2.5 W is sometimes insufficient to power external hard drives, technology is advancing and external drives gradually need less power, exacerbating the eSATA disadvantage. eSATAp (power over eSATA; aka ESATA/USB) is a connector introduced in 2009 that supplies power to attached devices using a new, backwards-compatible, connector. On a notebook eSATAp usually supplies only 5 V to power a 2.5 in HDD/SSD; on a desktop workstation it can additionally supply 12 V to power larger devices including 3.5 in HDD/SSD and 5.25 in optical drives.

eSATAp support can be added to a desktop machine in the form of a bracket connecting to motherboard SATA, power, and USB resources.

eSATA, like USB, supports hot plugging, although this might be limited by OS drivers and device firmware.

Thunderbolt

Thunderbolt combines PCI Express and DisplayPort into a new serial data interface that can be carried over longer and less costly cables. Thunderbolt has twice the transfer speed of USB 3.0 over copper wire.

Related standards

The PictBridge standard allows for interconnecting consumer imaging devices. It typically uses USB for its underlying communication layer.

The USB Implementers Forum is working on a wireless networking standard based on the USB protocol. Wireless USB is intended as a cable-replacement technology, and will use ultra-wideband wireless technology for data rates of up to 480 Mbit/s.

USB 2.0 High Speed Inter Chip (HSIC) is a chip-to-chip variant of USB 2.0 that eliminates the conventional analog transceivers found in normal USB. It was adopted as a standard by the USB Implementers Forum in 2007. The HSIC physical layer uses about 50% less power and 75% less board area compared to traditional USB 2.0. HSIC uses two signals at 1.2 V and has a throughput of 480 Mbit/s using 240 MHz DDR signaling. Maximum PCB trace length for HSIC is 10 cm. It does not have low enough latency to support RAM memory sharing between two chips.[75] [76]

See also

- Computer bus
- List of computer peripheral bus bit rates

Further reading

- *USB Complete : The Developer's Guide*; 4th Ed; Jan Axelson; 506 pages; 2009; ISBN 9781931448086.
- *USB Design by Example : A Practical Guide to Building I/O Devices*; 2nd Ed; John Hyde; 510 pages; 2001; ISBN 9780970284655.
- *USB Mass Storage : Designing and Programming Devices and Embedded Hosts*; 1st Ed; Jan Axelson; 287 pages; 2006; ISBN 9781931448048.
- *Serial Port Complete : COM Ports, USB Virtual COM Ports, and Ports for Embedded Systems*; 2nd Ed; Jan Axelson; 380 pages; 2007; ISBN 9781931448062.
- *Debugging USB 2.0 for Compliance: It's Not Just a Digital World: Agilent Technologies Application Note 1382-3*

References

[1] http://simson.net/clips/1999/99.Globe.05-20.USB_deserves_more_support+.shtml

[2] "SuperSpeed USB 3.0: More Details Emerge" (http://www.pcworld.com/article/156494/superspeed_usb_30_more_details_emerge.html). 6 Jan 2009. .

[3] "Definition of Universal Serial Bus" (http://books.google.com/books?id=fRvbxgH4wmsC&pg=PA7). *1394 Newsletter* **2** (4): 7–9. April 1998. . Retrieved 2010-03-11.

[4] Seebach, Peter (April 26, 2005). "Standards and specs: The ins and outs of USB" (http://www.ibm.com/developerworks/power/library/pa-spec7.html). IBM. . Retrieved 2010-03-11.

[5] *Universal Serial Bus Specification Revision 3.0 : 3.1* (http://www.usb.org/developers/docs/usb_30_spec_092911.zip). 9 September 2011. p. 41 (3-1). . Retrieved 14 October 2011.

[6] *Universal Serial Bus Specification Revision 3.0 : 1.6.* 9 September 2011. p. 31 (1–3).

[7] "USB 3.0 SuperSpeed gone wild at CES 2010, trumps even your new SSD" (http://www.engadget.com/2010/01/09/usb-3-0-superspeed-gone-wild-at-ces-2010-trumps-even-your-new-s/). January 9, 2010. . Retrieved 2011-02-20.

[8] "USB 3.0 Finally Arrives" (http://www.pcworld.com/article/186566/usb_30_finally_arrives.html). January 11, 2010. . Retrieved 2011-02-20.

[9] http://www.USB.org/

[10] "Usb.org" (http://www.usb.org/press/USB-IF_Press_Releases/2008_11_17_USB_IF.pdf) (PDF). . Retrieved 2010-06-22.

[11] *Universal Serial Bus Specification Revision 2.0* (http://www.usb.org/developers/docs/http://www.usb.org/developers/docs/usb_20_101111.zip). 11 October 2011. p. 13;30;256. . Retrieved 14 October 2011.

[12] *Universal Serial Bus Specification Revision 3.0 : 8.8.* 9 September 2011. pp. 8–25.

[13] USB Class Codes (http://www.usb.org/developers/defined_class) at USB.org

[14] Use class information in the interface descriptors. This base class is defined to be used in device descriptors to indicate that class information should be determined from the Interface Descriptors in the device.

[15] *Universal Serial Bus Test and Measurement Class Specification (USBTMC), Revision 1.0, April 14, 2003*, USB Implementers Forum, Inc.

[16] "DFU_1.1.doc" (http://www.usb.org/developers/devclass_docs/DFU_1.1.pdf) (PDF). . Retrieved 2010-06-22.

[17] *Universal Serial Bus 3.0 Specification*,4.4.11 "Efficiency"

[18] "100 Portable Apps for your USB Stick (both for Mac and Win)" (http://www.makeuseof.com/tag/portable-software-usb/). . Retrieved 2008-10-30.

[19] "Skype VoIP USB Installation Guide" (http://www.VoIP-Download.com/Skype.htm#USB/). . Retrieved 2008-10-30.

[20] "Universal Serial Bus Micro-USB Cables and Connectors Specification" (http://193.219.66.80/datasheets/usb_20/Micro-USB_final/Micro-USB_1_01.pdf) (PDF). USB Implementers Forum. 2007-04-04. . Retrieved 2010-09-03.

[21] *Universal Serial Bus 3.0 Specification: Revision 1.0.* June 6, 2011. pp. 531.

[22] "What is the Life Cycle of a USB Flash Drive?", GetUSB.info. Retrieved June 14, 2010. (http://www.getusb.info/what-is-the-life-cycle-of-a-usb-flash-drive/)

[23] Quinnell, Richard A (1996). "USB: a neat package with a few loose ends" (http://www.edn.com/archives/1996/102496/df_01.htm#USB fundamentals). *USB Fundamentals*. EDN Magazine of Reed Properties Inc. . Retrieved 2008-08-06.

[24] "Deprecation of the Mini-A and Mini-AB Connectors" (http://www.usb.org/developers/Deprecation_Announcement_052507.pdf) (PDF) (Press release). USB Implementers Forum. 2007-05-27. . Retrieved 2009-01-13.

[25] "ID Pin Resistance on Mini B-plugs and Micro B-plugs Increased to 1 Mohm" (http://compliance.usb.org/index.asp?UpdateFile=Cables and Connectors&Format=Standard#63). USB IF Compliance Updates. December 2009. . Retrieved 2010-03-01.

[26] "Mobile phones to adopt new, smaller USB connector" (http://www.usb.org/press/pressroom/2007_01_04_usbif.pdf) (PDF) (Press release). USB Implementers Forum. 2007-01-04. . Retrieved 2007-01-08.

[27] "Deprecation of the Mini-A and Mini-AB Connectors" (http://www.usb.org/developers/Deprecation_Announcement_052507.pdf) (PDF) (Press release). USB Implementers Forum. 2007-05-23. . Retrieved 2010-12-23.

[28] "Universal Serial Bus Micro-USB Cables and Connectors Specification to the USB 2.0 Specification, Revision 1.01" (http://193.219.66.80/datasheets/usb_20/Micro-USB_final/Micro-USB_1_01.pdf). USB Implementers Forum, Inc.. 2007-04-07. Archived from the original (http://www.usb.org/developers/docs/usb_20_101111.zip) on 2007-04-08. . Retrieved 2010-11-18. "Section 1.3: Additional requirements for a more rugged connector that will have durability past 10,000 cycles and still meet the USB 2.0 specification for mechanical and electrical performance was also a consideration. The Mini-USB could not be modified and remain backward compatible to the existing connector as defined in the USB OTG specification."

[29] "OMTP Local Connectivity: Data Connectivity" (http://www.omtp.org/Publications/Display.aspx?Id=08d2e4e3-ebee-407c-a51b-94057e7f7b19). Open Mobile Terminal Platform. 17 September 2007. . Retrieved 2009-02-11.

[30] http://europa.eu/rapid/pressReleasesAction.do?reference=MEMO/09/301; The following 10 biggest mobile phone companies have signed the MoU: Apple, LG, Motorola, NEC, Nokia, Qualcomm, Research In Motion, Samsung, Sony Ericsson, Texas Instruments

[31] "Commission welcomes new EU standards for common mobile phone charger" (http://europa.eu/rapid/pressReleasesAction.do?reference=IP/10/1776&format=HTML&aged=0&language=EN&guiLanguage=en). *Press Releases*. EUROPA. 2010-12-29. . Retrieved 2011-05-22.

[32] pressinfo (2009-10-22). "Press Release: Universal phone charger standard approved—One-size-fits-all solution will dramatically cut waste and GHG emissions" (http://www.itu.int/newsroom/press_releases/2009/49.html). Itu.int. . Retrieved 2009-11-04.

[33] http://www.instructables.com/id/XBox-360-Memory-Unit-USB-Connector/step3/Add-the-voltage-regulator/

[34] What is the USB 3.0 Cable Difference (http://hantat.com/News-Read-459-1.html)

[35] USB Frequently Asked Questions (http://www.usb.org/developers/usbfaq/#cab1) at USB.org. Retrieved 2010-12-10.

[36] "Propagation Delay" (http://www.wildpackets.com/support/compendium/ethernet/propagation_delay). . Retrieved 2008-10-31.

[37] Axelson, Jan. "USB 3.0 Developers FAQ" (http://www.lvr.com/usb3faq.htm#ca_maximum). . Retrieved 2011-10-29.

[38] "7.3.2 Bus Timing/Electrical Characteristics" (http://www.usb.org/developers/docs/). *Universal Serial Bus Specification*. USB.org. .

[39] "USB.org" (http://www.usb.org/developers/docs/). USB.org. . Retrieved 2010-06-22.

[40] "Roundup: 2.5-inch Hard Disk Drives with 500 GB, 640 GB and 750 GB Storage Capacities (page 17)" (http://www.xbitlabs.com/articles/storage/display/25inch-500-640-750gb-hdd-roundup_17.html#sect0). xbitlabs.com. 2010-06-16. . Retrieved 2010-07-09.

[41] "I have the drive plugged in but I cannot find the drive in "My Computer", why?" (http://www.simpletech.com/support/faq/portable_hard_disk_drives.php#1). simpletech.com. . Retrieved 2010-07-09.

[42] "Battery Charging Specification" (http://www.usb.org/developers/devclass_docs/batt_charging_1_1.zip). USB Implementers Forum, Inc. 15 April 2009. . Retrieved 23 September 2009.

[43] Minty Boost: The mysteries of Apple device charging (http://www.ladyada.net/make/mintyboost/icharge.html). Ladyada.net, (2010, 2011).

[44] "Usb-core.co.uk" (http://web.archive.org/web/20080614031146/http://www.usb-core.co.uk/03-04-2008-toshiba-announces-sleep-and-charge-usb-ports.html). Usb-core.co.uk. 2008-04-03. Archived from the original (http://www.usb-core.co.uk/03-04-2008-toshiba-announces-sleep-and-charge-usb-ports.html) on June 14, 2008. . Retrieved 2010-06-22.

[45] Cai Yan (2007-05-31). "China to enforce universal cell phone charger" (http://www.eetimes.com/rss/showArticle.jhtml?articleID=199800238&cid=RSSfeed_eetimes_newsRSS). EETimes.com. . Retrieved 2007-08-25.

[46] The Chinese FCC's technical standard: YD/T 1591-2006, "Technical Requirements and Test Method of Charger and Interface for Mobile Telecommunication Terminal Equipment". (http://www.dianyuan.com/bbs/u/63/2015571206841181.pdf) (Chinese)

[47] Crystal LAM Yan Yan and Harry LIU Yao Hui, ON Semiconductor. "How to conform to China's new mobile phone interface standards" (http://www.wirelessnetdesignline.com/202800278?printableArticle=true). Wireless Net DesignLine. . Retrieved 2010-06-22.

[48] "Pros seem to outdo cons in new phone charger standard" (http://www.news.com/2100-1041_3-6209247.html). news.com. September 20, 2007. . Retrieved 2007-11-26.

[49] "Press Release: Broad Manufacturer Agreement Gives Universal Phone Cable Green Light" (http://www.omtp.org/News/Display.aspx?Id=4ec69ecb-0978-4df6-b045-34557aabbcbd). OTMP. September 17, 2007. . Retrieved 2007-11-26.

[50] "GSM World agreement on Mobile phone Standard Charger" (http://www.gsmworld.com/newsroom/press-releases/2009/2548.htm). .

[51] "Common Charging and Local Data Connectivity" (http://www.omtp.org/Publications/Display.aspx?Id=4dda105f-8472-4c12-ba04-75dd3c1d4ca6). Open Mobile Terminal Platform. 11 February 2009. . Retrieved 2009-02-11.

[52] "Universal Charging Solution ~ GSM World" (http://www.gsmworld.com/our-work/mobile_planet/universal_charging_solution.htm). Gsmworld.com. . Retrieved 2010-06-22.

[53] "Meeting the challenge of the universal charge standard in mobile phones" (http://www.planetanalog.com/article/printableArticle.jhtml?articleID=218501515). PlanetAnalog.com. . Retrieved 2010-06-22.

[54] "CTIA–The Wireless Association Announces One Universal Charger Solution to Celebrate Earth Day" (http://www.ctia.org/media/press/body.cfm/prid/1817). Ctia.org. 2009-04-22. . Retrieved 2010-06-22.

[55] "Ec.europa.eu" (http://ec.europa.eu/enterprise/rtte/chargers.htm). Ec.europa.eu. 2009-06-29. . Retrieved 2010-06-22.

[56] Previous post Next post (2009-06-13). "Wired.com" (http://www.wired.com/gadgetlab/2009/06/europe-gets-universal-cellphone-charger-in-2010/). Wired.com. . Retrieved 2010-06-22.

[57] pressinfo (2009-10-22). "Itu.int" (http://www.itu.int/newsroom/press_releases/2009/49.html). Itu.int. . Retrieved 2010-06-22.
[58] http://www.macobserver.com/tmo/article/watt_to_know_about_iphone_ipad_power_adapters/
[59] "NEC ready to sample 'world's first' USB 3.0 controller chip" (http://www.reghardware.co.uk/2009/05/19/nec_usb_3_host/). . Retrieved 2009-06-15.
[60] "When will USB 3.0 products hit the market?" (http://www.everythingusb.com/superspeed-usb.html#6). . Retrieved 2009-05-11.
[61] "USB in a NutShell—Chapter 2—Hardware" (http://www.beyondlogic.org/usbnutshell/usb2.htm). Beyond Logic.org. . Retrieved 2007-08-25.
[62] "Technical Specifications of the USB 3.0 SuperSpeed Cables" (http://www.usb3.com/images/usb_superspeed_cable_spec.jpg). . 100717 usb3.com
[63] "Universal Serial Bus 3.0 Specification, Rev 1.0 November 12, 2008" (http://www.usb3.com/whitepapers/USB 3 0 (11132008)-final. pdf). . 100717 usb3.com
[64] "Seagate FreeAgent GoFlex Ultra-portable Review" (http://reviews.cnet.com/external-hard-drives/seagate-freeagent-goflex-ultra/4505-3190_7-34183942-2.html). Reviews.cnet.com. . Retrieved 2011-05-22.
[65] Real Deal (http://www.pcworld.com/article/82005/news_and_trends_usb_20s_real_deal.html).
[66] "FireWire vs. USB 2.0" (http://www.qimaging.com/support/pdfs/firewire_usb_technote.pdf) (PDF). QImaging. . Retrieved 20 July 2010.
[67] "FireWire vs. USB 2.0 - Bandwidth Tests" (http://www.cwol.com/firewire/firewire-vs-usb.htm). . Retrieved 2007-08-25.
[68] "USB 2.0 vs FireWire" (http://www.digit-life.com/articles/usb20vsfirewire). Digit-Life. . Retrieved 2007-08-25.
[69] Metz, Cade. "The Great Interface-Off: FireWire Vs. USB 2.0" (http://www.pcmag.com/article2/0,4149,847716,00.asp). PC Magazine. . Retrieved 2007-08-25.
[70] Heron, Robert. "USB 2.0 Versus FireWire" (http://www.g4tv.com/techtvvault/features/39129/USB_20_Versus_FireWire_pg3.html). TechTV. . Retrieved 2007-08-25.
[71] "FireWire vs. USB 2.0" (http://www.usb-ware.com/firewire-vs-usb.htm). USB Ware. . Retrieved 2007-03-19.
[72] Key, Gary (2005-11-15). "Firewire and USB Performance" (http://www.anandtech.com/mb/showdoc.aspx?i=2602&p=15). . Retrieved 2008-02-01.
[73] "Ieee 802.3" (http://standards.ieee.org/getieee802/download/802.3-2008_section1.pdf). IEEE. .
[74] By consumerist.com on December 18, 2006 7:32 PM (2010-03-08). "Powerbook Explodes After Comcast Plugs In Wrong Cable" (http://consumerist.com/2006/12/powerbook-explodes-after-comcast-plugs-in-wrong-cable.html). Consumerist.com. . Retrieved 2010-06-22.
[75] "Interchip Connectivity: HSIC, UniPro, HSI, C2C, LLI...oh my!" (http://info.arteris.com/blog/bid/59433/Interchip-Connectivity-HSIC-UniPro-HSI-C2C-LLI-oh-my). . Retrieved 24 June 2011.
[76] "USB High Speed Inter-Chip Interface" (http://www.interfacebus.com/hsic-bus-high-speed-inter-chip-usb.html). . Retrieved 24 June 2011.

External links

- USB official website (USB Implementers Forum, Inc.) (http://www.usb.org/)
- Intel Universal Host Controller Interface (UHCI) (http://download.intel.com/technology/usb/UHCI11D.pdf)
- USB 3.0 Standard-A, Standard-B, Powered-B connectors pinouts (http://pinoutsguide.com/Slots/usb_3_0_connector_pinout.shtml)
- Agilent characterization and compliance test (http://www.agilent.com/find/USB)

Secure Digital

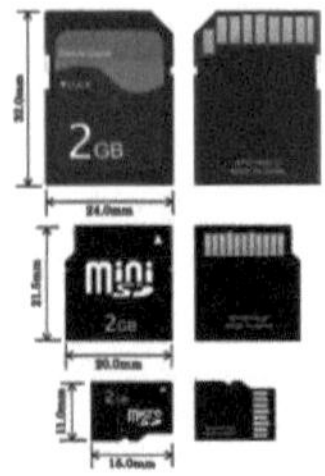

SD (top), miniSD, microSD cards

Media type	Memory card
Capacity	SDSC (SD): 1 MB to 2 GB although 4 GB cards are available SDHC: 4 GB to 32 GB SDXC: >32 GB to 2 TB
Developed by	SD Card Association
Dimensions	Standard: 32×24×2.1 mm Mini: 21.5×20×1.4 mm Micro: 15×11×1.0 mm
Weight	Standard: ~2.0 g Mini: ~1.0 g Micro: ~0.5 g
Usage	Portable devices, including digital cameras and handheld computers
Extended from	MultiMediaCard (MMC)

Secure Digital (**SD**) is a non-volatile memory card format developed by the SD Card Association (SDA) for use in portable devices. The SD technology is used by more than 400 brands across dozens of product categories and more than 8,000 models, and is considered the de-facto industry standard.[1]

SD now comprises several families of cards:[2] the original, Standard-Capacity (SDSC) card, a High-Capacity (SDHC) card family, an eXtended-Capacity (SDXC) card family,[3] and the SDIO[4] family with input/output functions rather than just data storage.

SD also comprises three different form factors: the original size, the "mini" size, and the "micro" size (see illustration). Electrically passive adaptors allow the use of a smaller card in a host device built to hold a larger card. There are many combinations of form factors and device families.

Host devices that comply with newer versions of the specification provide backward compatibility and accept older SD cards, but older host devices do not recognize newer cards. The SDA uses several trademarked logos to enforce compliance with its specifications and assure users of compatibility.[5] This article explains several factors that can prevent the use of a newer SD card:

- A newer card may offer greater capacity than the host device can handle
- A newer card may use a file system the host device cannot navigate
- Use of an SDIO card requires the host device be designed for the input/output functions the card provides
- The organization of the card was changed starting with the SDHC family
- Some vendors produced SDSC cards above 1 GB before the SDA had standardized a method of doing so.

Types of cards

The SDA extended the SD specification in various ways:

- It defined electrically identical cards in smaller sizes: miniSD and microSD (originally named TransFlash). Smaller cards are usable in larger slots through use of a passive adapter. By comparison, Reduced Size MultiMediaCards (RS-MMCs) are simply shorter MMCs and can be used in MMC slots by use of a physical extender.
- It defined higher-capacity cards, some with faster speeds and added capabilities: SDHC (Secure Digital High Capacity) and SDXC (Secure Digital eXtended Capacity). These cards redefine the interface so that they cannot be used in older host devices.
- It defined an SDIO card family that provides input-output functions and may also provide memory functions. These cards are only fully functional in host devices designed to support their input-output functions.

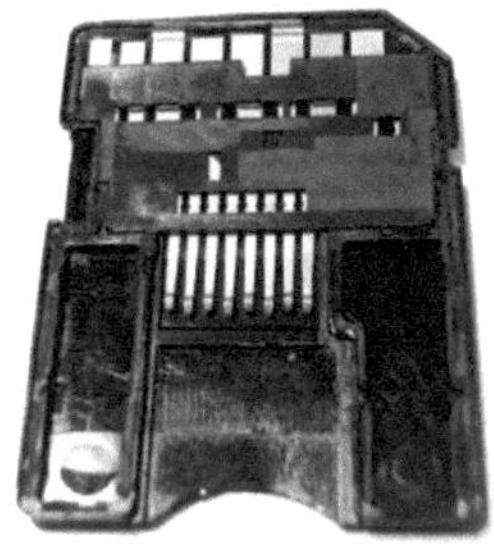

Dismantled microSD to SD adapter showing the passive connection from the microSD card slot on the bottom to the SD pins on the top

Physical size

The SD card specification defines three physical sizes. The SD and SDHC families are available in all three sizes, but the SDXC family is not available in the mini size, and the SDIO family is not available in the micro size.

Standard size

- SD, SDHC, SDXC, SDIO
- 32 mm × 24 mm × 2.1 mm
- 32 mm × 24 mm × 1.4 mm (as thin as MMC) for **Thin SD** (rare)

Mini size

- miniSD, miniSDHC, miniSDIO
- 21.5 mm × 20 mm × 1.4 mm

Micro size

- microSD, microSDHC, microSDXC

Size comparison of families: SD, miniSD, microSD

- 15 mm × 11 mm × 1.0 mm

SDHC

The Secure Digital High Capacity (SDHC) format is defined in Version 2.0 of the SD specification. It supports cards with capacities up to 32 GB.[1] The SDHC trademark is licensed to ensure compatibility.[6]

16 GB SDHC Card

SDHC cards are physically and electrically identical to standard-capacity SD cards (SDSC). The major compatibility issues between SDHC and SDSC cards, as covered above, are the redefinition of the Card-Specific Data (CSD) register in Version 2.0, and the fact that SDHC cards are typically shipped preformatted with the FAT32 file system.

Host devices that accept SDHC cards are required to accept SDSC cards.[1] However, host devices designed for SDSC do not recognize SDHC memory cards, although some devices can do so through a firmware upgrade. Microsoft Windows may need a hotfix to support accessing SDHC cards.[7] [8] [9]

SDXC

The Secure Digital Extended Capacity (SDXC) format supports cards up to 2 TB (2048 GB), compared to a limit of 32 GB for SDHC cards in the SD 2.0 specification.

History

SDXC was announced at Consumer Electronics Show (CES) 2009 (January 7–10, 2009). At the same show, SanDisk and Sony also announced a comparable Memory Stick XC variant with the same 2 TB maximum as SDXC,[10] and Panasonic announced plans to produce 64 GB SDXC cards.[11]

On March 6, 2009, Pretec introduced the first SDXC card,[12] a 32 GB card with a read/write speed of 400 Mbit/s. But it was not until early 2010 that compatible host devices came onto the market, including Sony's Handycam HDR-CX55V camcorder, Canon's EOS Rebel T2i Digital SLR camera,[13] a USB card reader from Panasonic, and an integrated SDXC card reader from JMicron.[14] The earliest laptops to integrate SDXC card readers relied on a USB 2.0 bus, which does not have the bandwidth to support SDXC at full speed.[15]

Also in early 2010, commercial SDXC cards appeared from Toshiba (64 GB),[16] [17] Panasonic (64 GB and 48 GB),[18] and SanDisk (64 GB).[19] In early 2011, Centon Electronics, Inc. (64 GB and 128 GB) and Lexar (128 GB)[20] began shipping SDXC cards rated at Speed Class 10. Pretec offered cards from 8 GB to 128 GB rated at Speed Class 16.[21]

In September 2011, SanDisk released the first microSDXC card (64 GB).[22]

Compatibility with SDHC

SDXC host devices will accept all previous families of SD memory cards. Conversely, SDHC host devices will accept SDXC cards that follow Version 3.0, since the interface is identical,[3] but the following issues may affect usability:

- SDXC cards that follow Version 4.0 will use the exFAT file system, which the host device might not support.
- An SDHC host device will not test the capability bits newly defined for SDXC cards. It will therefore not be able to use the new features of SDXC, such as transfer speeds above UHS104.[23]

The maximum capacity of 32 GB for SDHC cards put them all within the limitations imposed by the FAT32 file system. Full support for all SDXC cards will require that the host device support exFAT.

As of 2011, Windows versions that support SDXC are: Microsoft Windows 7, Windows Vista SP1+,[3] Windows XP SP2 or SP3 with KB955704,[24] Windows Server 2008 SP1+, Windows Server 2003 SP2 or SP3 with KB955704, Windows CE 6+, Mac OS X 10.6.5 or later.[25]

Linux systems that support SDHC cards also support SDXC cards, as long as the card contains a compatible file system. A proprietary driver is available for the exFAT file system.[26]

SDIO

A SDIO (Secure Digital Input Output) card is an extension of the SD specification to cover I/O functions. Host devices that support SDIO (typically PDAs like the Palm Treo, but occasionally laptops or mobile phones) can use the SD slot to support GPS receivers, modems, barcode readers, FM radio tuners, TV tuners, RFID readers, digital cameras, and interfaces to Wi-Fi, Bluetooth, Ethernet, and IrDA. Many other SDIO devices have been proposed, but it is now more common for I/O devices to connect using the USB interface.

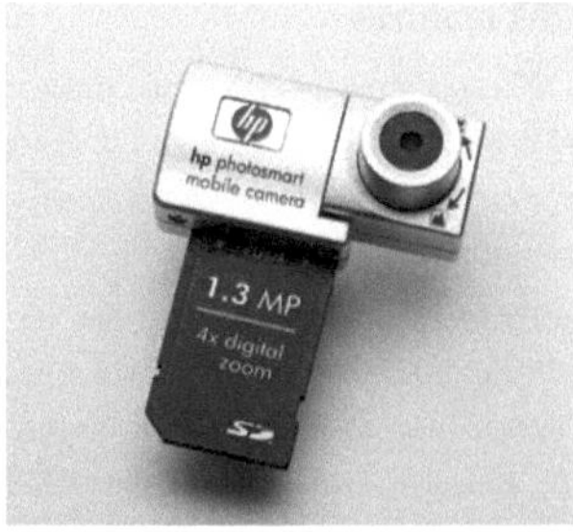

Camera using the SDIO interface to connect to some HP iPAQ devices

SDIO cards support most of the memory commands of SD cards. SDIO cards can be structured as 8 logical cards, although currently, the typical way that an SDIO card uses this capability is to structure itself as one I/O card and one memory card.

Host support for SDIO

The SDIO and SD interfaces are mechanically and electrically identical. Host devices built for SDIO cards generally accept SD memory cards without I/O functions. However, the reverse is not true, because host devices need suitable drivers and applications to support the card's I/O functions. For example, an HP SDIO camera usually does not work with PDAs that do not list it as an accessory. Inserting an SDIO card into any SD slot causes no physical damage nor disruption to the host device, but users may be frustrated that the SDIO card does not function fully when inserted into a seemingly compatible slot. (Bluetooth devices exhibit comparable compatibility issues, although to a lesser extent thanks to standardized Bluetooth profiles.)

Features

Card security (non-DRM)

Write protection

When looking at the card from the top, the right side (the side with the beveled corner) must be notched.

On the left side, there may be a write-protection notch. If the notch is omitted, the card can be read and written. If the card is notched, it is read-only, except that the notch may be partly covered by a sliding tab. In this case, the user can slide the tab upward (toward the contacts) to declare the card read/write, or downward to declare it read-only.

The presence of a notch, and the presence and position of a tab, have no effect on the SD card's operation. A host device that supports write protection should refuse to write to an SD card that is designated read-only in this way. Some host devices do not support write protection, which is an optional feature of the SD specification. Host devices that do obey a read-only indication may give the user a way to override it. The miniSD and microSD formats do not support write protection with the notch-and-tab method.

Typical cards sold as a medium for protected content are permanently marked read-only by having a notch and no sliding tab.

Card password

A host device can lock an SD card using a password of up to 16 bytes, typically supplied by the user. A locked card interacts normally with the host device except that it rejects commands to read and write data. A locked card can be unlocked only by providing the same password. The host device can, after supplying the old password, specify a new password or disable locking. Without the password (typically, in the case that the user forgets the password), the host device can command the card to erase all the data on the card for future re-use (except card data under DRM), but there is no way to gain access to the existing data.

DRM features

All SD cards incorporate a digital rights management (DRM) scheme. Roughly 10% of the storage capacity of an SD card is not available to the user, but is used by the on-card processor to verify the identity of an application program that it will then allow to read protected content. The card prohibits other accesses, such as users trying to make copies of protected files.

The DRM scheme embedded in the SD cards is the Content Protection for Recordable Media (CPRM or CPPM) specification of the 4C Entity, which features the Cryptomeria cipher (also termed *C2*). The specification is kept secret and is accessible only to licensees. The scheme has not been broken or hacked, but this feature of SD cards is rarely actually used to protect content. DVD-Audio uses the same DRM scheme.

Windows Media files can be DRM-encoded so as to make use of the SD card's DRM abilities.

Windows Phone 7 devices use SD cards designed to be accessed only by the phone manufacturer or mobile provider. An SD card inserted into the phone underneath the battery compartment becomes locked "to the phone with an automatically generated key" so that "the SD card cannot be read by another phone, device, or PC".[27] Symbian devices, however, are some of the very few which can perform the necessary low-level format operations on locked SD cards. It is therefore possible to use a device such as the Nokia N8 to reformat the card for subsequent use in other devices.[28]

The *Super Digital* cards manufactured by Super*Talent are the same in appearance and function as Secure Digital cards, but they lack the CPRM feature of Secure Digital cards.[29]

Vendor enhancements

Vendors have sought to differentiate their products in the market through various vendor-specific features:

SD cards with dual-interface SD and USB connections

- **Integrated Wi-Fi** — Eye-Fi produces several SD cards with built-in Wi-Fi transceivers providing static security (WEP 40; 104; and 128, WPA-PSK, and WPA2-PSK). The card lets any digital camera with an SD slot transmit captured images over a wireless network, or store the images on the card's memory until it is in range of a wireless network.[30] Some models geotag their pictures.
- **Pre-loaded content** — In 2006, SanDisk announced Gruvi, a microSD card with extra Digital Rights Management (DRM) features, with which to use memory cards as a medium to sell content. SanDisk again announced pre-loaded cards in 2008, under the slotMusic name, this time not using any of the DRM capabilities of the SD card.[31] In 2011, SanDisk offered various collections of 1000 songs on a single slotMusic card for about $40,[32] now restricted to compatible devices and without the ability to copy the files.
- **Integrated USB connector** — The SanDisk **SD Plus** product can be plugged directly into a USB port without needing a USB card reader.[33] Other companies introduced comparable

products, such as the **Duo SD** product of OCZ Technology and the **3 Way** (microSDHC, SDHC, and USB) product of A-DATA, which was available in 2008 only.

- **Different colors** — SanDisk has used various colors of plastic or adhesive label, including a "gaming" line in translucent plastic colors that indicated the card's capacity.
- **Integrated display** — In 2006, A-DATA announced an SD card with a digital display that would show how much free space was left on the card.[34] [35]

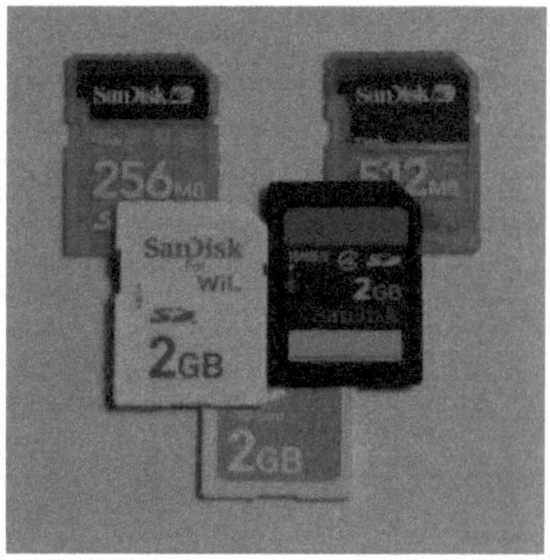

SanDisk cards in various colors

Speeds

An SD card's speed is measured by how quickly an amount of information can be read from, or written to, the card. In applications that require sustained write throughput, such as video recording, the device might not perform satisfactorily if the SD card's class rating falls below a particular speed. For example, a camcorder built for a Class 6 card may suffer dropouts or corrupted video if a slower card is used. Digital cameras may experience a noticeable lag between shots, while the camera writes the picture to a slower card.

A card's speed depends on many factors, such as the following:

- The likelihood of soft errors that the card's controller must re-try
- The fact that, on most cards, writing data requires the controller to read and erase a larger region, then rewrite that entire region with the desired part changed
- The possibility of fragmentation: that a body of information the host views as a unit is, for historical reasons, written to non-contiguous regions of memory. (This possibility does not cause rotational or head-movement delays as with magnetic media, but it does vary the amount of computation the card's controller must do.)

In early SD cards, the speed was measured with the × rating, which compared the average speed of reading data to that of the original CD-ROM drive. Currently, the official unit of measurement is the **Speed Class Rating**, which guarantees a minimum rate at which data can be written to the card.

The newer families of SD card improve card speed by increasing the bus rate (the frequency of the clock signal that strobes information into and out of the card). Whatever the bus rate, the card can signal to the host that it is "busy" until a read or a write operation is complete. Compliance with a higher speed rating is a guarantee that the card limits its use of the "busy" indication.

Speed Class Rating

The **Speed Class Rating** is the official unit of speed measurement for SD cards. The class number guarantees a minimum write speed as a multiple of 8 Mbit/s (1 MB/s). The SD Association defines several speed class ratings, but manufacturers may claim conformance to those ratings without independent verification.

32GB SDHC card

Unlike the earlier "×" speed ratings, the host device can read a card's speed class. A device can warn the user if the card reports a speed class that falls below an application's minimum need.[36]

These are the ratings of all currently available cards:[37] [38]

Class	Speed
Class 2	2 MB/s
Class 4	4 MB/s
Class 6	6 MB/s
Class 10	10 MB/s

Speed Classes 2, 4, and 6 assert that the card supports the respective number of MB/s as a minimum sustained write speed for a card in a fragmented state. Class 10 asserts that the card supports 10 MB/s as a minimum non-fragmented sequential write speed.[36] By comparison, the older "×" rating measured maximum speed under ideal conditions, and was vague as to whether this was read speed or write speed.

× rating

The × rating is a multiple of the standard CD-ROM drive speed of 1.2 Mbit/s (approximately 150 kB/s). Basic cards transfer data up to six times (6×) the CD-ROM speed; that is, 7.2 Mbit/s. The 2.0 specification defines speeds up to 200×, but is not as specific as Speed Classes are on how to measure speed. Manufacturers may report best-case speeds and may report the card's fastest read speed, which is typically faster than the write speed. Vendors including Transcend and Kingston report their cards' write speed.[39]

Card with 150x rating

This table lists common ratings, the minimum transfer rates, and the corresponding Speed Class (though the comparison is not always exact).

Rating	Read Speed (MB/s)	Write Speed (MB/s)	Speed Class
6×	0.9		
10×	1.5		
13×	**2.0**	**2.0**	**2**
26×	**4.0**	**4.0**	**4**
32×	4.8	5.0	
40×	**6.0**	**6.0**	**6**
66×	**10.0**	**10.0**	**10**
100×	15.0	15.0	
133×	20.0	20.0	
150×	22.5	22.5	
200×	30.0	30.0	
266×	40.0	40.0	
300×	45.0	45.0	
400×	60.0	60.0	
600×	90.0	90.0	

UHS Speed Class

The Ultra-High Speed (UHS) interface is available on some SDHC and SDXC cards.[40] The following ultra-high speeds are specified:

- **UHS-I** cards, specified in SD Version 3.01,[36] support a clock frequency of 100 MHz (a quadrupling of the original Default Speed (DS), which in four-bit transfer mode could transfer 50 MB/s. UHS-I cards declared as **UHS104** also support a clock frequency of 208 MHz, which could transfer 104 MB/s. UHS-I is the only class for which products are currently available.[38]
- Double data rate operation at 50 MHz (DDR50) is also specified in Version 3.01, and is mandatory for microSDHC and microSDXC cards labeled as UHS-I. In this mode, four bits are transferred when the clock signal rises and another four bits when it falls, transferring an entire byte on each full clock cycle.
- **UHS-II** cards, to be defined in Version 4.0, further raise the data transfer rate to a theoretical maximum of 312 MB/s.[41]

UHS memory cards work best with UHS host devices. The combination lets the user record HD resolution videos to tapeless camcorders while performing other functions. It is also suitable for real-time broadcasts and capturing large HD videos.

Cards that comply with UHS show UHS-I or UHS-II on the label, and report this capability to the host device. Use of UHS requires that the host device command the card to drop from 3.3-volt to 1.8-volt operation and select the 4-bit transfer mode.

Market penetration

Secure Digital cards are ubiquitous in consumer electronic devices, and have become the dominant means of storing several gigabytes of data in a small size.

Devices such as netbooks, digital cameras, camcorders, personal digital assistant (PDAs), mobile phones, video game consoles, digital audio players, and many others use them.

Smaller devices tend to use microSD or miniSD rather than full sized SD cards.

SD cards are not generally used in mass produced devices where only a small amount of storage is needed due to economic reasons, or where a very large amount of storage is needed.

Digital cameras

A camcorder with a 4 GB SDHC card

SD/MMC cards replaced Toshiba's SmartMedia as the dominant memory card format used in digital cameras. In 2001, SmartMedia had achieved nearly 50% use, but by 2005 SD/MMC had achieved over 40% of the digital camera market and SmartMedia's share had plummeted, with cards not being easily available in 2007.

At this time all the leading digital camera manufacturers use SD in their consumer product lines, including Canon, Casio, Fujifilm, Kodak, Leica, Nikon, Olympus, Panasonic, Pentax, Ricoh, Samsung, and Sony. Formerly, Olympus and Fujifilm used XD-Picture Cards (xD cards) exclusively, while Sony only used Memory Stick; however as of January 2010, all three support SD.

Some prosumer and professional digital camera models continue to offer CompactFlash, either on a second card slot or as the only storage, as it has historically offered a better price/capacity ratio and faster transfer rates.

Secure Digital memory cards can be used in Sony XDCAM EX camcorders via the MEAD-SD01 adapter.[42]

Personal computers

Although many personal computers accommodate SD cards as an auxiliary storage device through a built-in slot or a USB adaptor, SD cards cannot be used as the primary hard disk through the onboard ATA controller because none of the SD card variants supports ATA signalling. This use requires a separate SD controller chip[43] or a SD-to-CompactFlash converter. SD cards can be used to boot and run an Operating System using a USB interface on a computer that supports booting from USB; in this scenario the Operating System must support USB access immediately in the post boot environment.

Embedded systems

In 2008, the SD Card Association specified Embedded SD, "leverag[ing] well-known SD standards" to enable non-removable SD-style devices on printed circuit boards.[44] SanDisk provides such memory components under the iNAND brand.[45]

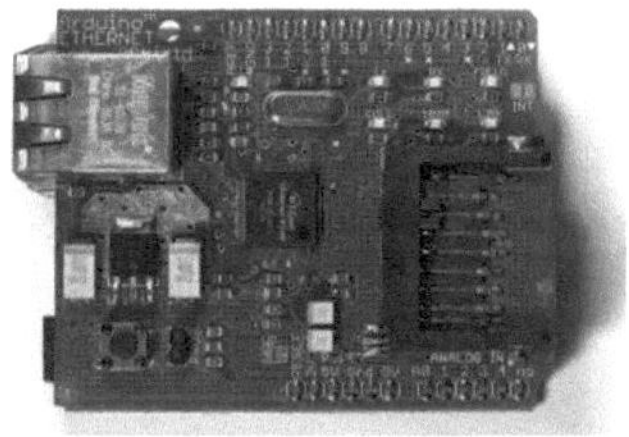

This shield (daughterboard) gives Arduino prototyping microprocessors access to SD cards plugged into the socket.

Most modern microcontrollers have built-in SPI logic that can interface to a SD card operating in its SPI mode, providing non-volatile storage. Even if a microcontroller lacks the SPI feature, the feature can be emulated by bit banging. For example, a home-brew hack combines spare General Purpose Input/Output (GPIO) pins of the processor of the Linksys WRT54G router with MMC support code from the Linux kernel.[46] This technique can achieve throughput of up to 1.6 Mbit/s.

History

In 1999, SanDisk, Matsushita, and Toshiba agreed to develop and market the Secure Digital (SD) Memory Card, which was a development of the MultiMediaCard (MMC). The new card provided both digital rights management (DRM) up to the Secure Digital Music Initiative (SDMI) standard, and a high memory density for the time.

On top: a microSDHC card that stores about 8 billion bytes. Below it: 64 units of magnetic-core memory, used until the 1970s, that stored eight bytes.

The new format was designed to compete with the Memory Stick, a DRM product that Sony released the prior year. It was mistakenly predicted that DRM features[47] would be widely used due to pressure from music and other media suppliers to prevent piracy.

The trademarked *SD* logo was originally developed for the Super Density Disc, which was the unsuccessful Toshiba entry in the DVD format war. This is why the *D* resembles an optical disc.

At the 2000 Consumer Electronics Show (CES) trade show, the three companies announced the creation of the SD Card Association to promote SD cards. The association's headquarters are in California and it comprises some 30 companies that produce devices and content. Early samples of the SD Card were available in the first quarter of 2000, with production quantities of 32 and 64 MB cards available 3 months later.

Relation to MMC

The SD cards changed the MMC design in several ways:

- Asymmetrical slots in the sides of the SD card prevent inserting it upside down, while an MMC will go in most of the way but make no contact if inverted.
- Most SD cards are 2.1 mm thick, compared to 1.4 mm for MMCs. The SD specification defines a card called **Thin SD** with a thickness of 1.4 mm, but they are rare, as the SDA went on to define even smaller form factors.
- The card's electrical contacts are recessed beneath the surface of the card, protecting them from contact with a user's fingers.

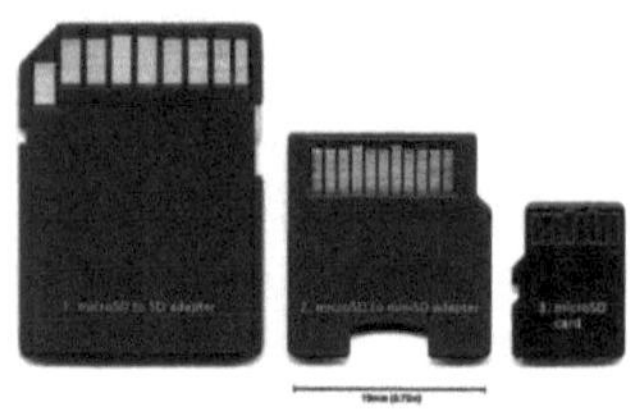

microSD to SD adapter (left), microSD to miniSD adapter (middle), microSD card (right)

- The SD specification envisaged capacities and transfer rates exceeding those of MMC, and these have both grown over time. For a comparison table, see below.

Full-sized SD cards will not fit into the slimmer MMC slots, and there are other issues that affect the ability to use one format in a host device designed for the other.

Follow-on products

In March 2003, SanDisk Corporation announced the introduction of the miniSD and demonstrated it at CeBIT 2003.[48] The miniSD card was adopted in 2003 by the SD Association [49] as a small form factor extension to the SD card standard. While the new cards were designed especially for use in mobile phones, they are usually packaged with a miniSD adapter which enables compatibility with all devices equipped with a standard SD memory card slot.

At CTIA Wireless 2005, the SDA announced the small **microSD** form factor (and SDHC, with capacities in excess of 2 GB and a minimum sustained read and write speed of 17.6 Mbit/s). SanDisk had conceived microSD when its CTO and the CTO of Motorola concluded that current memory cards were too large for mobile phones. The card was originally called T-Flash, but just before product launch, T-Mobile sent a cease and desist order to SanDisk claiming that T-Mobile owned the trademark on T-(anything), and the name was changed to TransFlash. TransFlash and microSD cards are the same; each can be used in devices made for the other.[50] SanDisk induced the SDA to administer the microSD standard. The SDA approved the final microSD specification on July 13, 2005. Initially, microSD cards were available in capacities of 32, 64, and 128 MB.

In April 2006, the SD Association released a detailed specification for the non-security related parts of the SD memory card standard and for the Secure Digital Input Output (SDIO) cards and the standard SD host controller.

In September, 2006, SanDisk announced the 4 GB miniSDHC.[51] Like the SD and SDHC, the miniSDHC card has the same form factor as the older miniSD card but the HC card requires HC support built into the host device. Devices that support miniSDHC will work with miniSD and miniSDHC, but devices without specific support for miniSDHC will work only with the older miniSD card.

In January 2009, the SD Association announced the SDXC family that will support cards up to 2 TB memory size and speeds up to 300 Mbyte/s.[52]

The miniSD and microSD cards can be used directly in SD slots with a simple passive adapter, since the cards differ in size and shape but not electrical interface. With an active electronic adapter, SD cards can be used in CompactFlash or PC card slots. Some SD cards include a USB connector for compatibility with desktop and laptop computers, and card readers allow SD cards to be accessed via connectivity ports such as USB, FireWire, and the parallel printer port. SD cards can also be accessed via a floppy disk drive with a FlashPath adapter.

Openness of the standard

Like most memory card formats, SD is covered by numerous patents and trademarks. Royalties for SD card licences are imposed for manufacture and sale of memory cards and host adapters (USD$1,000/year plus membership at USD$1,500/year), but SDIO cards can be made without royalties.

Size comparison of various flash cards: SD, CompactFlash, MMC, xD

Early versions of the SD specification were available only after agreeing to a non-disclosure agreement (NDA) that prohibited development of an open source driver. However, the system was eventually reverse-engineered, and free software drivers provided access to SD cards that did not use DRM. Since then, the (SDA) has provided a simplified version of the specification under a less restrictive license.[53] Although most open-source drivers were written before this, it has helped them solve compatibility issues.

In 2006, the SDA also released a simplified version of the specification of the host controller interface (as opposed to the specification of SD cards) and later also for the physical layer, ASSD extensions, SDIO, and SDIO Bluetooth Type-A, under a disclaimers agreement.[54] Again, most of the information had already been discovered and Linux had a fully free driver for it. Still, building a chip conforming to this specification caused the One Laptop per Child project to claim "the first truly Open Source SD implementation, with no need to obtain an SDI license or sign NDAs to create SD drivers or applications."[55]

USB-based Memory card reader that supports multiple flash card types

The fact that the complete SD specification is proprietary mainly affects embedded systems and laptops, since users of desktop PCs generally read SD cards via USB-based card readers. These card readers present a standard USB mass storage interface to memory cards, thus separating the operating system from the details of the underlying SD interface. However, embedded systems (such as portable music players) usually gain direct access to SD cards and thus need complete programming information. Desktop card readers are themselves embedded systems; their manufacturers have usually paid the SDA for complete access to the SD specifications. Many notebook computers now include SD card readers not based on USB; device drivers for these essentially gain direct access to the SD card, as do embedded systems.

Technical details

Transfer modes

Various SD cards may support various combinations of the following bus types and transfer modes. The SPI bus and one-bit SD bus are mandatory for all SD families, as explained in the next section.

- **SPI:** Serial Peripheral Interface Bus is primarily used by embedded microcontrollers. This bus type supports only a 3.3-volt interface.
- **One-bit SD:** Separate command and data channels and a proprietary transfer format.
- **Four-bit SD:** Uses extra pins plus some reassigned pins. UHS-I and UHS-II requires this bus type.

Interface

Command interface

SD cards and host devices initially communicate through a synchronous one-bit interface, where the host device provides a clock signal that strobes single bits into and out of the SD card. The host device thereby sends 48-bit commands and receives responses. The card can signal that a response will be delayed, but the host device can abort the dialogue.

Through issuing various commands, the host device can:

- Determine the type, memory capacity, and capabilities of the SD card
- Command the card to use a different voltage, different clock speed, or advanced electrical interface
- Prepare the card to receive a block to write to the flash memory, or read and reply with the contents of a specified block.

The command interface is an extension of the MultiMediaCard (MMC) interface. SD cards dropped support for some of the commands in the MMC protocol, but added commands related to copy protection. By using only commands supported by both standards until determining the type of card inserted, a host device can accommodate both SD and MMC cards.

Electrical interface

SD card pin assignment

All SD card families initially use a 3.3-volt electrical interface. On command, SDHC and SDXC cards switch to 1.8-volt operation.[36]

At initial power-up or card insertion, the host device selects either the Serial Peripheral Interface (SPI) bus or the one-bit SD bus by the voltage level present on Pin 1. Thereafter, the host device may issue a command to switch to the four-bit SD bus interface, if the SD card supports it. For various card types, support for the four-bit SD bus is either optional or mandatory.[36]

After determining that the SD card supports it, the host device can also command the SD card to switch to a higher transfer speed. Until determining the card's capabilities, the host device should not use a clock speed faster than 400 KHz. SD cards other than SDIO (see below) have a Default Speed (DS) clock rate of up to 25 MHz. The host device is not required to use the maximum clock speed that the card supports. It may conserve power to operate at less than the maximum clock speed.[36] Between commands, the host device can stop the clock entirely.

SDIO cards

A special family of SD cards, called SDIO, comprises Low-Speed and Full-Speed cards. Both types of SDIO cards support SPI and one-bit SD bus types. Low-Speed SDIO cards are allowed to also support the four-bit SD bus; Full-Speed SDIO cards are required to support the four-bit SD bus. To use a SDIO card as a "combo card" (for both memory and I/O), the host device must first select four-bit SD bus operation. Two other unique features of Low-Speed SDIO are a maximum clock rate of 400 kHz for all communications, and the use of Pin 8 as "interrupt" to try to initiate dialogue with the host device.[56]

SD card pin-outs

SPI Bus

Pin	Name	I/O	Logic	Description
1	nCS	I	PP	Card Select (Neg True)
2	DI	I	PP	Data In [MOSI]
3	VSS	S	S	Ground
4	VDD	S	S	Power
5	CLK	I	PP	Clock [SCLK]
6	VSS	S	S	Ground
7	DO	O	PP	Data Out [MISO]
8	NC nIRQ	. O	. OD	NC (Memory Cards) Interrupt (SDIO Cards)
9	NC	.	.	NC

One-Bit SD Bus

Pin	Name	I/O	Logic	Description
1	NC	.	.	NC
2	CMD	I/O	PP,OD	Command, Response
3	VSS	S	S	Ground
4	VDD	S	S	Power
5	CLK	I	PP	Clock
6	VSS	S	S	Ground
7	DAT0	I/O	PP	Data 0
8	NC nIRQ	. O	. OD	NC (Memory Cards) Interrupt (SDIO Cards)
9	NC	.	.	NC

Four-Bit SD Bus

Pin	Name	I/O	Logic	Description
1	DAT3	I/O	PP	Data 3
2	CMD	I/O	PP,OD	Command, Response
3	VSS	S	S	Ground
4	VDD	S	S	Power
5	CLK	I	PP	Clock
6	VSS	S	S	Ground
7	DAT0	I/O	PP	Data 0
8	DAT1 nIRQ	I/O O	PP OD	Data 1. SDIO Cards share with Interrupt Period
9	DAT2	I/O	PP	Data 2

Notes:

1. Direction is relative to card. I = Input, O = Output.
2. PP = Push-Pull logic, OD = Open-Drain logic.
3. S = Power Supply, NC = Not Connected (or logical high).

Ganging cards together

The one-bit SD protocol was derived from the MMC protocol, which envisaged the ability to put up to 3 cards on a bus of common signal lines. The cards use open collector interfaces, where a card may pull a line to the low voltage level; the line is at the high voltage level (because of a pull-up resistor) if no card pulls it low. Though the cards shared clock and signal lines, each card had its own chip select line to sense that the host device had selected it.

The SD protocol envisaged the ability to gang 30 cards together without separate chip select lines. The host device would broadcast commands to all cards and identify the card to respond to the command using its unique serial number.

In practice, cards are rarely ganged together because open-collector operation has problems at high speeds and increases power consumption. Newer versions of the SD specification recommend separate lines to each card.

Achieving higher card speeds

The SD specification defines four-bit-wide transfers. (The MMC specification supports this and also defines an eight-bit-wide mode.) Transferring several bits on each clock pulse improves the card speed. Advanced SD families have also improved speed by offering faster clock frequencies and double data rate (explained here).

File system

Like other types of flash memory card, an SD card of any SD family is a block-addressable storage device, in which the host device can read or write fixed-size blocks by specifying their block number.

MBR and FAT

Most SD cards ship preformatted with one or more MBR partitions, where the first or only partition contains a file system. This lets them operate like the hard disk of a personal computer. Typically, an SD card is formatted with MBR and the following file system:

- For SDSC cards: FAT16
- For SDHC cards: FAT32
- For SDXC cards:
    - Early SDXC cards follow the SD 3.0 specification, although use of FAT32 prevents support for the full 2 TB maximum capacity it specifies.[23]
    - Version 4.0 of the SD specification selected Microsoft's proprietary exFAT file system, which will enable higher-capacity and faster cards.[57]

Most consumer products that take an SD card will expect it to be partitioned and formatted in this way. The wide support for FAT16 and FAT32 lets these cards be used on virtually any host device with an SD reader, to present the user with the familiar method of named files in a hierarchical directory tree.

On such SD cards, standard utility programs such as SCANDISK can be used to repair or retrieve corrupted data, and sometimes recover deleted files. Defragmentation tools for FAT file systems may be used on such cards. The resulting consolidation of files may provide a marginal improvement in the time required to read or write the file,[37] but not an improvement comparable to defragmentation of hard drives, where storing a file in multiple fragments may involve a time penalty to move between physical areas of the drive. Moreover, defragmentation performs writes to the SD card that count against the card's rated lifespan. The write endurance of the physical memory is discussed in the article on flash memory; newer technology to increase the storage capacity of a card currently provides worse write endurance.

When reformatting an SD card smaller than 4 GB, FAT16 should be used. (This is also an option for 4 GB cards, but it requires the use of 64 kiB clusters, which are not widely supported.) FAT16 does not support cards above 4 GB.

Other file systems

Because the host views the SD card as a block storage device, the card does not require MBR partitions or any specific file system. The card can be reformatted to use any file system the operating system supports. For example:

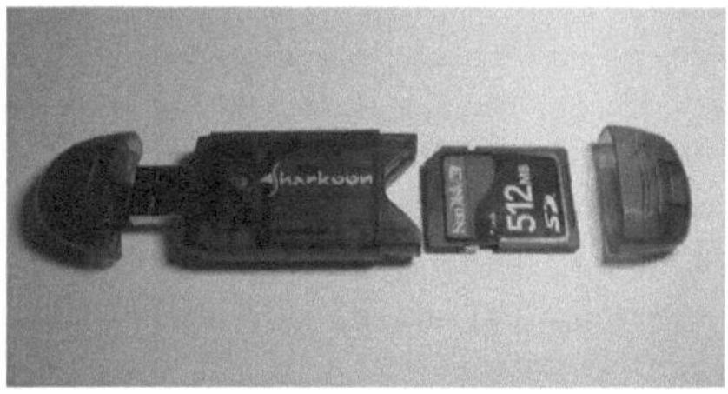

USB-based SD card reader

- Under Unix-like operating systems such as Linux or FreeBSD, SD cards can be formatted using the UFS, EXT3, ext4, btrFS, HFS Plus, or the ReiserFS file systems.
- Under Mac OS X, SD cards can be partitioned as GUID devices and formatted with the HFS Plus file system.
- Under Windows and some Unix systems, SD cards can be formatted using NTFS and, on later versions, exFAT.

Additionally, an SD card called Live SD can contain an embedded operating system (such as Live USB). Computers that can bootstrap from an SD card (either using a USB adapter or inserted into the computer's flash media reader) instead of the hard disk drive may thereby be able to recover from a corrupted hard disk drive. A Live SD can be write-locked to preserve the system's integrity.

Risks of reformatting

Reformatting an SD card with a different file system, or even with the same one, may make the card slower, or shorten its lifespan. Some cards use wear leveling, in which frequently modified blocks are mapped to different portions of memory at different times, and some wear-leveling algorithms are designed for the access patterns typical of the file allocation table on a FAT16 or FAT32 device.[58] In addition, the preformatted file system may use a cluster size that matches the erase region of the physical memory on the card; reformatting may change the cluster size and make writes less efficient.

Power use

The power consumption of microSD cards varies by manufacturer, but appears to be in the range of 66-330 mW (20-100 mA at a supply voltage of 3.3 V). Specifications from TwinMos technologies list a maximum of 149 mW (45 mA) during transfer.[59] Toshiba, on the other hand, lists 264-330 mW (80-100 mA).[60]

Storage capacity and incompatibilities

All SD cards let the host device determine how much information the card can hold, and the specification of each SD family gives the host device a guarantee of the maximum capacity a compliant card will report.

By the time the Version 2.0 (SDHC) specification was completed in June 2006,[61] vendors had already devised 2 GB and 4 GB SD cards, either as specified in Version 1.1, or by creatively reading it, or by violating it but remaining in its framework. The resulting cards do not work correctly in some host devices.[62] [63]

SDSC cards above 1 GB

4 GB SDSC card

A host device can ask any inserted SD card for its 128-bit identification string (the Card-Specific Data or CSD). In standard-capacity cards (SDSC), 12 bits identify the number of memory clusters (ranging from 1 to 4,096) and 3 bits identify the number of blocks per cluster (which decode to 4, 8, 16, 32, 64, 128, 256, or 512 blocks per cluster). The host device multiplies these figures (as shown in the following section) with the number of bytes per block to determine the card's capacity in bytes.

In SD version 1.00, the number of bytes per block was assumed to be 512. This permitted SDSC cards up to 4,096 × 512 × 512 = 1 GB, for which there are no known incompatibilities.

Version 1.01 let an SDSC card use a 4-bit field to indicate 1,024 or 2,048 bytes per block instead.[36] Doing so enabled cards with 2 GB and 4 GB capacity.

Early SDSC host devices that assume 512-byte blocks therefore will not fully support the insertion of 2 GB or 4 GB cards. In some cases, the host device can read data that happens to reside in the first 1 GB of the card. If the assumption is made in the driver software, success may differ from one version of Windows to another. In addition, any host device might not support a 4 GB SDSC card, since the specification lets it assume that 2 GB is the maximum for these cards.

Storage capacity calculations

The format of the Card-Specific Data (CSD) register changed between version 1 (SDSC) and version 2.0 (which defines SDHC and SDXC).

Version 1

In Version 1 of the SD specification, capacity is calculated by combining fields of the CSD as follows:

```
Capacity=(C_SIZE+1)<<(C_SIZE_MULT+2)<<READ_BL_LEN        2 GiB max.
Where 0<=C_SIZE<=4095, 0<=C_SIZE_MULT<=7, READ_BL_LEN==9 || READ_BL_LEN==10
```

Later versions state (at Section 4.3.2) that a 2 GB SDSC card shall set its READ_BL_LEN (and WRITE_BL_LEN) to indicate 1024 bytes, so that the above computation correctly reports the card's capacity; but that, for consistency, the host device shall not request (by CMD16) block lengths over 512 bytes.[36]

Versions 2 and 3

In the definition of SDHC and SDXC cards in Version 2.0, the C_SIZE portion of the CSD is 22 bits and it indicates the memory size in multiples of 512 KB. SDHC cards are restricted to reporting a capacity not over 32 GiB. (The C_SIZE_MULT field is removed and READ_BL_LEN is no longer used to compute capacity.) Two bits that were formerly reserved now identify the card family: 0 is SDSC; 1 is SDHC or SDXC; 2 and 3 are reserved.[36] Because of these redefinitions, older host devices do not correctly identify SDHC or SDXC cards nor their correct capacity.

Capacity is calculated thus:

```
Capacity=(C_SIZE+1)*524288
where for SDHC       4112<=C_SIZE<=65375       (approx. 2 GB) < capacity < 32 GiB
      for SDXC      65535<=C_SIZE                      32 GiB <= capacity <= 2 TiB max.
```

Capacities above 4 GB can only be achieved by following Version 2.0 or later versions. In addition, capacities equal to 4 GB must also do so to guarantee compatibility.

Version 4

Version 4.0 lets SDXC cards use all 22 bits of the C_SIZE field. An SDHC card that did so (reported C_SIZE > 65375 to indicate a capacity of over 32 GiB) would violate the specification. A host device that relied on C_SIZE rather than the specification to determine the card's maximum capacity might support such a card, but the card might fail in other SDHC-compatible host devices.

Comparison to other flash memory formats

Overall, SD is less open than CompactFlash or USB flash memory drives; these are open standards which can be implemented free of payment for licensing, royalties, or documentation. (CompactFlash and USB flash drives may, however, require licensing fees for the use of the SDA's trademarked logos.)

However, SD is much more open than Memory Stick, for which no public documentation nor any documented legacy implementation is available. All SD cards can be accessed freely using the well-documented SPI bus.

xD cards are simply 18-pin NAND flash chips in a special package and support the standard command set for raw NAND flash access. Although the raw hardware interface to xD cards is well understood, the layout of its memory contents—necessary for interoperability with xD card readers and digital cameras—is totally undocumented. The consortium that licenses xD cards has not released any technical information to the public.

Type	MMC	RS-MMC	MMC Plus	SecureMMC	SD	SDIO	miniSD	microSD
SD Socket	Yes	Mechanical adapter	Yes	Yes	Yes	Yes	Electromechanical adapter	Electromechanical adapter
Pins	7	7	13	7	9	9	11	8
Form factor	shallow	shallow/narrow	shallow	shallow	deep (some)	deep	narrow/slim/shallow	narrow/slim/extra shallow
Breadth	24 mm	24 mm	24 mm	24 mm	24 mm	24 mm	20 mm	11 mm
Width	32 mm	18 mm	32 mm	32 mm	32 mm	32 mm+	21.5 mm	15 mm
Depth	1.4 mm	1.4 mm	1.4 mm	1.4 mm	2.1 mm (some)	2.1 mm	1.4 mm	1 mm
SPI mode	Optional	Optional	Optional	Yes	Yes	Yes	Yes	Yes
1-bit mode	Yes	Yes	Yes	Yes	Yes	Yes	Yes	Yes
4-bit mode	No	No	Yes	?	Optional	Optional	Optional	Optional
8-bit mode	No	No	Yes	No	No	No	No	No
Interrupts	No	No	No	No	No	Optional	No	No
Max clock rate	20 MHz	20 MHz	52 MHz	20 MHz?	208 MHz	50 MHz	208 MHz	208 MHz
Max transfer	20 Mbit/s	20 Mbit/s	416 Mbit/s	20 Mbit/s?	832 Mbit/s	200 Mbit/s	832 Mbit/s	832 Mbit/s
Max SPI transfer	20 Mbit/s	20 Mbit/s	52 Mbit/s	20 Mbit/s	50 Mbit/s	50 Mbit/s	50 Mbit/s	50 Mbit/s
DRM	No	No	No	Yes	Yes	N/A	Yes	Yes
User encrypt	No	No	No	Yes	No	No	No	No
Simplified spec	Yes	Yes	No	Not yet?	Yes	Yes	No	No
Membership cost	JEDEC: $4400/yr, optional				SD Card Association: $2000/yr, general; $4500/yr, executive			

Specification cost	Free			?	Simplified spec: free Full spec: free to members, $1000/yr to R&D non-members			
Host license	No	No	No	No	Yes $1000/yr			
Card royalties	Yes	Yes	Yes	Yes	Yes	Yes +$1000/yr	Yes	Yes
Open source compatible	Yes	Yes	Yes?	Yes?	Yes	Yes	Yes	Yes
Nominal operating voltage	3.3V	1.8V/3.3V	1.8V/3.3V[64] [65]	1.8V/3.3V	3.3V	3.3V	3.3V	3.3V
Type	**MMC**	**RS-MMC**	**MMC Plus**	**SecureMMC**	**SD**	**SDIO**	**miniSD**	**microSD**

Table data compiled mostly from simplified versions of MMC and SDIO specifications and other data on SD card and MMC association web sites. Data for other card variations is interpolated.

Capacity limit in all SD/MMC formats appears to be 128 GB in LBA mode (28-bit sector address).

See also

- SD Card Association
- Comparison of memory cards
- Serial Peripheral Interface Bus (SPI)
- File Allocation Table (FAT16, FAT32) and exFAT
- Flash memory
- slotMusic
- MultiMediaCard
- USB FlashCard
- Universal Flash Storage

References

Inside a 512 MB SD card: NAND flash chip that holds the data (bottom) and SD controller (top)

[1] "About" (http://www.sdcard.org/developers/about/). SD Card Association. . Retrieved 2011-12-08.

[2] "Capacity" (http://www.sdcard.org/developers/overview/capacity/). SD Card Association. . Retrieved 2011-12-08.

[3] "Using SDXC" (http://www.sdcard.org/consumers/sdxc_capabilities/using_sdxc/). SD Card Association. . Retrieved 2011-12-08.

[4] "SDIO" (http://www.sdcard.org/developers/overview/sdio/). SD Card Association. . Retrieved 2011-12-08.

[5] "Using SD Memory Cards is Easy" (video on YouTube) - SD Card Association (http://www.youtube.com/watch?v=wX6NwBa1csY)

[6] What are SDHC, miniSDHC, and microSDHC? (http://www.sandisk.com/Assets/File/pdf/retail/SDHC1.pdf) SanDisk.com

[7] "Microsoft Support 934428 - Hotfix for Windows XP that adds support for SDHC cards that have a capacity of more than 4 GB" (http://support.microsoft.com/kb/934428). Support.microsoft.com. 2008-02-15. . Retrieved 2010-08-22.

[8] "Microsoft Support 939772 - Some Secure Digital (SD) cards may not be recognized in Windows Vista" (http://support.microsoft.com/kb/939772). Support.microsoft.com. 2008-05-15. . Retrieved 2010-08-22.

[9] "Microsoft Support 949126 - A Secure Digital High Capacity (SDHC) card is not recognized on a Windows Vista Service Pack 1-based computer" (http://support.microsoft.com/kb/949126). Support.microsoft.com. 2008-02-21. . Retrieved 2010-08-22.

[10] "SanDisk and Sony to expand Memory Stick Pro and Memory Stick Micro formats" (http://www.sandisk.com/about-sandisk/press-room/press-releases/2009/2009-01-07-sandisk-and-sony-to-expand-âmemory-stick-proâ-and-âmemory-stick-microâ-formats). Sandisk.com. . Retrieved 2010-08-22.
[11] "SD Card, Memory Stick formats to reach 2 terabytes, but when? - Betanews" (http://www.betanews.com/article/SD_Card_Memory_Stick_formats_to_reach_2_terabytes_but_when/1231453659). . 090108 betanews.com
[12] "Pretec introduces world's first SDXC card: Digital Photography Review" (http://www.dpreview.com/news/0903/09030601pretecsdxc.asp). Dpreview.com. 2009-03-06. . Retrieved 2010-08-22.
[13] "Canon EOS Rebel T2i / 550D Digital SLR Camera Review" (http://www.the-digital-picture.com/Press-Release/Canon-EOS-Rebel-T2i-550D-Digital-SLR-Camera-Press-Release.aspx). . The-Digital-Picture.com
[14] Ng, Jansen (2009-11-24). "Lack of Card Readers Holding Back SDXC Flash Memory Adoption" (http://www.dailytech.com/Lack+of+Card+Readers+Holding+Back+SDXC+Flash+Memory+Adoption/article16915.htm). DailyTech. . Retrieved 2009-12-22.
[15] Ng, Jansen (2009-11-30). "Lenovo, HP, Dell Integrating SDXC Readers in New 32nm Intel "Arrandale" Laptops" (http://www.dailytech.com/Lenovo+HP+Dell+Integrating+SDXC+Readers+in+New+32nm+Intel+Arrandale+Laptops/article16937.htm). DailyTech. . Retrieved 2009-12-22.
[16] Ng, Jansen (2009-12-22). "Toshiba Sampling First SDXC Flash Memory Cards" (http://www.dailytech.com/Toshiba+Sampling+First+SDXC+Flash+Memory+Cards/article16972.htm). DailyTech. . Retrieved 2009-12-22.
[17] "Toshiba's 64 GB SDXC card to finally go on sale (in Japan)" (http://www.crunchgear.com/2010/04/15/toshibas-64gb-sdxc-card-to-finally-go-on-sale-in-japan/). CrunchGear. . Retrieved 2010-08-09.
[18] "Panasonic Introduces New 64 GB* and 48 GB* SDXC Memory Cards, Available Globally in February 2010" (http://www2.panasonic.com/webapp/wcs/stores/servlet/prModelDetail?storeId=11301&catalogId=13251&itemId=389511&modelNo=Content01052010041118461&surfModel=Content01052010041118461). Panasonic. . Retrieved 2010-08-09.
[19] "Sandisk ships its highest capacity sd card ever" (http://sandisk.com/about-sandisk/press-room/press-releases/2010/2010-02-22-sandisk-ships-its-highest-capacity-sd-card-ever,-the-64gb-sandisk-ultra-sdxc-card). SanDisk. . Retrieved 2010-08-09.

Inside a 2 GB SD card: two NAND flash chips (top and middle), SD controller chip (bottom)

Inside a SDHC card

[20] Lexar ships 128 GB Class 10 SDXC card; March 2011. (http://www.betanews.com/article/Lexar-ships-first-128GB-SDXC-cards/1300305310)
[21] "SDXC/SDHC 433X Class 16 Card from Pretec" (http://www.pretec.com/news-event/press-room/item/root/sdxcsdhc-433x-c16). Pretec. 2011-06-13. . Retrieved 2010-12-03.
[22] First 64GB microSD Card Here; When Will Smartphones Support It? (http://pocketnow.com/smartphone-news/first-64gb-microsd-card-arrives-when-will-smartphones-support)
[23] "SD Association Triples Speeds with UHS-II" (https://www.sdcard.org/home/SD_Association_Announces_UHS-II_eBOOK_Jan_5_2011_ENGLISH.PDF). . Retrieved 2011-08-09.
[24] http://www.microsoft.com/downloads/details.aspx?FamilyID=1cbe3906-ddd1-4ca2-b727-c2dff5e30f61&displaylang=en
[25] "About the SD and SDXC card slots" (http://support.apple.com/kb/HT3553). Apple Inc.. 2011-05-03. . Retrieved 2011-09-05.
[26] "Tuxera exFAT Embedded" (http://www.tuxera.com/products/tuxera-exfat-embedded/). Tuxera.com. 2010-06-01. . Retrieved 2011-07-11.
[27] "Windows Phone 7 Secure Digital Card Limitations" (http://support.microsoft.com/kb/2450831). .
[28] "Windows Phone 7's microSD mess: the full story (and how Nokia can help you out of it)" (http://www.engadget.com/2010/11/17/windows-phone-7s-microsd-mess-the-full-story-and-how-nokia-ca). .
[29] "Super Talent Technology - DDR and DDR2 Memory" (http://www.supertalent.com/products/sd.php). Supertalent.com. . Retrieved 2010-08-22.
[30] "Home" (http://www.eye.fi). Eye-Fi. . Retrieved 2010-08-22.
[31] AudioHolics (http://www.audioholics.com/news/industry-news/sandisk-slotmusic)
[32] "slotRadio" (http://www.sandisk.com/consumer-products/slotradio). SanDisk. . Retrieved 2011-11-27.
[33] http://www.theregister.co.uk/2005/07/25/review_sandisk_ultra_ii_sd_plus/

[34] "I4U News - A DATA Announces SD Card w/ Bi-stable Capacity Display" (http://www.i4u.com/article7106.html). I4u.com. 2006-11-14. . Retrieved 2010-08-22.
[35] OhGizmo! » Archive » Company Announces SD Card With Embedded Capacity Display (http://www.ohgizmo.com/2006/11/13/company-announces-sd-card-with-embedded-capacity-display/)
[36] "SD Part 1, Physical Layer Simplified Specification, Version 3.01" (http://www.sdcard.org/downloads/pls/simplified_specs/Part_1_Physical_Layer_Simplified_Specification_Ver_3.01_Final_100518.pdf). . Retrieved 2011-12-04.
[37] Fragmentation and Speed (https://www.sdcard.org/developers/overview/speed_class/) SDCard.org
[38] "SD Speed Class/UHS Speed Class" (https://www.sdcard.org/developers/overview/speed_class/). . Retrieved 21 Nov 2011.
[39] "Kingston Technology Company - Flash Memory Cards and X-Speed Ratings" (http://www.kingston.com/flash/x/default.asp). Kingston.com. . Retrieved 2010-08-22.
[40] "SD cards branded with an upper-case 'I' are faster, yo" (http://www.engadget.com/2010/06/24/sd-cards-branded-with-an-upper-case-i-are-faster-yo/). Engadget. . Retrieved 2010-08-22.
[41] "SD Association announces UHS-II, ultra high-speed SD card specification" (http://www.robgalbraith.com/bins/content_page.asp?cid=7-11133-11156). Rob Galbraith. 2011-01-05. . Retrieved 2011-01-05.
[42] MEAD-SD01 SDHC card adapter (Sony) (http://pro.sony.com/bbsc/ssr/micro-xdcamexsite/cat-accessories/product-MEADSD01/)
[43] "TS-7800 Embedded" (http://www.embeddedarm.com/products/board-detail.php?product=TS-7800). Embeddedarm.com. . Retrieved 2010-08-22.
[44] "Embedded SD" (http://www.sdcard.org/developers/overview/embedded_sd/). SD Card Association. . Retrieved 2011-11-30.
[45] "iNAND Embedded Flash Drives" (http://www.sandisk.com/business-solutions/inand-embedded-flash-drives). SanDisk. . Retrieved 2011-11-30.
[46] "Linksys WRT54G-TM SD/MMC mod - DD-WRT Wiki" (http://www.dd-wrt.com/wiki/index.php/Linksys_WRT54G-TM_SD/MMC_mod). Dd-wrt.com. 2010-02-22. . Retrieved 2010-08-22.
[47] "Press Releases 17 July 2003" (http://www.toshiba.co.jp/about/press/2003_07/pr1701.htm). Toshiba. 2003-07-17. . Retrieved 2010-08-22.
[48] SanDisk Introduces The World's Smallest Removable Flash Card For Mobile Phones-The miniSD Card (http://www.sandisk.com/Corporate/PressRoom/PressReleases/PressRelease.aspx?ID=1536) SanDisk.com
[49] http://www.sdcard.org/
[50] PhoneScoop - Sandisk T-Flash announcement. (http://www.phonescoop.com/news/item.php?n=801)
[51] SanDisk Introduces 4GB miniSDHC Flash Card for Mobile Phones (http://www.sandisk.com/Corporate/PressRoom/PressReleases/PressRelease.aspx?ID=3530) SanDisk.com
[52] SD Association - Press Release - Announced SDXC. (http://www.sdcard.org/press/SD_Association_Announces_SDXC_Revised_1-7-09.pdf)
[53] "Sharp Linux PDA promotes the use of proprietary SD card, but more open MMC works just fine" (http://www.linux.com/archive/feed/20060). Linux.com. . Retrieved 2010-08-22.
[54] Simplified Specification Agreement (http://www.sdcard.org/developers/tech/sdcard/pls/) from the SDA's website
[55] "OLPC mailing list archive" (http://mailman.laptop.org/pipermail/community-news/2006-September/000023.html). Mailman.laptop.org. . Retrieved 2010-08-22.
[56] "Simplified Version of SDIO CARD SPEC" (https://www.sdcard.org/developers/overview/sdio/sdio_spec/). SD Card Association. . Retrieved 2011-12-09.
[57] "SDXC memory cards promise 2TB of storage, 300MBps transfer" (http://www.engadget.com/2009/01/07/sdxc-memory-cards-promise-2tb-of-storage-300mbps-transfer/). Engadget.com. . Retrieved 2010-08-22.
[58] "Optimizing Linux with cheap flash drives" (http://lwn.net/Articles/428584/). Linux Weekly News. . Retrieved 2011-04-11.
[59] http://www.twinmos.com.tw
[60] http://www.toshiba-memory.com/en/micro_sd_cards.html
[61] A look into how SDHC will affect the future Nand Flash market (http://www.dramexchange.com/WeeklyResearch/Post/1/492.aspx). DRAMeXchange, December 2006
[62] SD Compatibility (http://www.hjreggel.net/cardspeed/special-sd.html), CARDSPEED - Card Readers and Memory Cards, December 1, 2006
[63] "WinXP SP3 can't read 4GB SD card in multicard reader" (http://www.eggheadcafe.com/software/aspnet/33338344/winxp-sp3-cant-read-4gb.aspx). Eggheadcafe.com. . Retrieved 2010-08-22.
[64] JEDEC MMC 4.4 Standard Pg.7 (http://www.jedec.org/download/search/JESD84-A44.pdf), http://www.jedec.org 2008
[65] Transcend v4.0 card does not support 1.8V (http://www.transcendusa.com/support/dlcenter/datasheet/TSxxMMC4.pdf), http://www.transcendusa.com 2009

External links

Organizations

- Official website (http://www.sdcard.org), SD Association, sdcard.org
  - Membership: $2000/yr for General, $4500/yr for Executive.
  - Full Specification: Free for members, $1000/yr for R&D non-members.

Specifications

- SD Simplified Specifications (https://www.sdcard.org/downloads/pls/), sdcard.org, Free
  - Part 1, Physical Layer, v3.01
  - Part A1, ASSD Extension, v2.00
  - Part A2, SD Host Controller, v3.00
  - Part E1, SDIO, v3.00
  - Part E2, SDIO Bluetooth Type-A, v1.00
- Microsoft Extensible Firmware Initiative FAT32 File System Specification (http://download.microsoft.com/download/1/6/1/161ba512-40e2-4cc9-843a-923143f3456c/fatgen103.doc), Microsoft

Software

- SD Formatter for SD / SDHC / SDXC cards (Windows and Mac) (https://www.sdcard.org/downloads/formatter_3/), sdcard.org

Comparisons

- Comparison of numerous memory cards and readers, plus technical information (http://www.hjreggel.net/cardspeed/index.html#special-sd.html), hjreggel.net
- SDHC Card comparison (german) (http://www.hardware-infos.com/tests.php?test=81), hardware-infos.com
- Speed Testing of UHS-1 Cards in prosumer Nikon D7000 Camera (http://glamourphotography.co/gear/uhs-speed-class-1-sdhc-memory-cards-tested-delkin-8gb-elite-633x-secure-digital-uhs-i-95-mbsec-vs-sandisk-extreme-pro-32gb-)
- Nikon D7000 Memory speed tests, SDHC Memory Cards: including Sandisk Extreme Pro UHS Speed Class 1 45 Mbyte/s and Several Popular Class 10 SDHC Cards (http://glamourphotography.co/gear/nikon-d7000-memory-speed-tests-sdhc-memory-cards-including-sandisk-extreme-pro-uhs-speed-class-1-45mbsec-and-several-pc)

Interfacing

- Interfacing to SD cards, great technical details (http://elm-chan.org/docs/mmc/mmc_e.html)
- Interfacing AVR (Arduino) to SD cards, C source code (http://www.dharmanitech.com/2009/01/sd-card-interfacing-with-atmega8-fat32.html)
- Interfacing ARM to SD cards, C source code (http://gandalf.arubi.uni-kl.de/avr_projects/arm_projects/arm_memcards/index.html)
- Interfacing MSP430 to SD cards, C source code (http://www.cs.ucr.edu/~amitra/sdcard/Additional/sdcard_appnote_foust.pdf), Michigan State University
- Interfacing MAXQ2000 to SD cards, good technical descriptions, C source code (http://www.maxim-ic.com/appnotes.cfm/an_pk/3969), maxim-ic.com
- SD card controller, Verilog source code (http://www.opencores.org/project,sdcard_mass_storage_controller), opencores.org

PictBridge

PictBridge is an industry standard introduced in 2003 from the Camera & Imaging Products Association (CIPA) for direct printing. It allows images to be printed directly from digital cameras to a printer, without having to connect the camera to a computer. Its formal name is "Standard of Camera & Imaging Products Association CIPA DC-001 — 2003 Digital Solutions for Imaging Devices" [1]. CIPA DC-001-2003 Rev. 2.0 has been published in 2007.

The PictBridge Logo

PictBridge is typically implemented using Universal Serial Bus ports and the USB protocol. PictBridge-capable printers typically have a USB Type A receptacle which is then connected by cable to the USB port of a PictBridge-capable digital camera (typically a Type Mini-B). The user selects the images on the camera, which the printer retrieves and prints.

Proprietary ownership

PictBridge arguably does not qualify as an open standard, as the specification can only be obtained from CIPA after agreement not to disclose any information from the specification to others (section 2.2 of the agreement [2]). In practice this means that PictBridge cannot be implemented as free software or open source software other than by reverse-engineering the protocol (perhaps aided by the white paper [3] that CIPA provides), if publishing source code of an implementation of the PictBridge standard is considered to count as "disclosing information" from the specification.

A printer may implement similar functions to a PictBridge printer without the non-disclosure agreement by treating the camera's memory as a USB mass storage device, although the user interface for image selection would necessarily be on the printer rather than the camera in this case.

See also

- Picture Transfer Protocol
- Digital Print Order format

External links

- CIPA PictBridge Standard Website [4]
- Explanation on Canon's website [5] at the Wayback Machine (*archived September 27, 2009*)

References

[1] http://www.cipa.jp/english/pictbridge/FAQ_E_05122003.pdf
[2] http://www.cipa.jp/pictbridge/documents_e/ObtainingStandard_E.pdf
[3] http://www.cipa.jp/pictbridge/documents_e/WhitePaperE_Rev.2.0.pdf
[4] http://www.cipa.jp/english/pictbridge/
[5] http://web.archive.org/20090927080404/http://www.canon.com.au/about/press_room/story_830.html

Audio Video Interleave

Filename extension	.avi
Internet media type	video/vnd.avi[1] video/avi video/msvideo video/x-msvideo
Type code	'Vfw '
Uniform Type Identifier	public.avi
Developed by	Microsoft
Container for	Audio, Video

Audio Video Interleave (also **Audio Video Interleaved**), known by its acronym **AVI**, is a multimedia container format introduced by Microsoft in November 1992 as part of its Video for Windows technology. AVI files can contain both audio and video data in a file container that allows synchronous audio-with-video playback. Like the DVD video format, AVI files support multiple streaming audio and video, although these features are seldom used. Most AVI files also use the file format extensions developed by the Matrox OpenDML group in February 1996. These files are supported by Microsoft, and are unofficially called "AVI 2.0".

Format

AVI is a derivative of the Resource Interchange File Format (RIFF), which divides a file's data into blocks, or "chunks." Each "chunk" is identified by a FourCC tag. An AVI file takes the form of a single chunk in a RIFF formatted file, which is then subdivided into two mandatory "chunks" and one optional "chunk".

The first sub-chunk is identified by the "hdrl" tag. This sub-chunk is the file header and contains metadata about the video, such as its width, height and frame rate. The second sub-chunk is identified by the "movi" tag. This chunk contains the actual audio/visual data that make up the AVI movie. The third optional sub-chunk is identified by the "idx1" tag which indexes the offsets of the data chunks within the file.

By way of the RIFF format, the audio-visual data contained in the "movi" chunk can be encoded or decoded by software called a codec, which is an abbreviation for (en)coder/decoder. Upon creation of the file, the codec translates between raw data and the (compressed) data format used inside the chunk. An AVI file may carry audio/visual data inside the chunks in virtually any compression scheme, including Full Frame (Uncompressed), Intel Real Time (Indeo), Cinepak, Motion JPEG, Editable MPEG, VDOWave, ClearVideo / RealVideo, QPEG, and MPEG-4 Video.

Metadata

As a derivative of the Resource Interchange File Format (RIFF), AVI files are commonly tagged with metadata in the INFO chunk. In addition, AVI files can embed Extensible Metadata Platform (XMP). By design, any RIFF file can legally include additional chunks of data, each identified by a four-character code; software which does not understand that particular code should simply skip the chunk. As such, it is theoretically possible to expand any RIFF file format, including AVI, to support almost any conceivable metadata. Some of the limitations of AVI in modern use relate to a lack of standardization in this metadata (see Limitations below).

Limitations

Since its introduction in the early 90s, new computer video techniques have been introduced which the original AVI specification did not anticipate:

- AVI does not provide a standardized way to encode aspect ratio information, with the result that players cannot select the right one automatically (though it may be possible to do so manually).[2]
- There are several competing approaches to including a time code in AVI files, which affects usability of the format in film and television post-production, although it is widely used. For WAV audio files, Broadcast Wave extensions were designed to standardize post-production metadata, but an equivalent for AVI files has not emerged.
- AVI is not intended to contain variable frame rate material. Workarounds for this limitation increase overhead dramatically.
- AVI was not intended to contain video using any compression technique which requires access to future video frame data beyond the current frame. Approaches exist to support modern video compression techniques (such as MPEG-4) which rely on this function, although this is beyond the intent of the original specification and may cause problems with playback software which does not anticipate this use.
- AVI cannot contain some specific types of variable bitrate (VBR) data reliably (such as MP3 audio at sample rates below 32 kHz).
- Overhead for AVI files at the resolutions and frame rates normally used to encode standard definition feature films is about 5 MB per hour of video, the significance of which varies with the application.

More recent container formats (such as Matroska, Ogg and MP4) solve all these problems, although software is freely available to both create and correctly replay AVI files which use these more recent techniques.

DV AVI

DV AVI (aka DV-AVI) is a type of AVI file where the video has been compressed to conform with DV standards. There are two types of DV-AVI files:

- Type 1: The multiplexed Audio-Video is kept in its original multiplexing and saved together into the Video section of the AVI file
    - Does not waste much space (audio is saved uncompressed, but even uncompressed audio is tiny compared to the video part of DV), but Windows applications based on the VfW API do not support it.
- Type 2: Like type 1, but audio is also saved as an additional audio stream into the file.
    - Supported by VfW applications, at the price of a small increase in file size.

Type 1 is actually the newer of the two types. Microsoft made the "type" designations, and decided to name their older VfW-compatible version "Type 2", which only furthered confusion about the two types. In the late 1990s through early 2000s, most professional-level DV software, including non-linear editing programs, only supported Type 1. One notable exception was Adobe Premiere, which only supported Type 2. High-end FireWire controllers usually captured to Type 1 only, while "consumer" level controllers usually captured to Type 2 only. Software is and was available for converting Type 1 AVIs to Type 2, and vice-versa, but this is a time-consuming process.

Many current FireWire controllers still only capture to one or the other type. However, almost all current DV software supports both Type 1 and Type 2 editing and rendering, including Adobe Premiere. Thus, many of today's users are unaware of the fact that there are two types of DV AVI files. In any event, the debate continues as to which – Type 1 or Type 2 – if either, is better.

See also

- Comparison of container formats
- List of video editing software

References

[1] Microsoft Corporation (June 1998). "WAVE and AVI Codec Registries - RFC 2361" (http://tools.ietf.org/html/rfc2361). IETF. . Retrieved 2009-12-06.

[2] "Determining the Shape of Pixels and Frames" (http://www.microsoft.com/windows/windowsmedia/howto/articles/PixelFrames.aspx). .

External links

- John P. McGowan's AVI Overview (http://www.jmcgowan.com/avi.html)
- Microsoft's AVI RIFF File Reference (http://msdn2.microsoft.com/en-us/library/ms779636.aspx)
- PVDTools' RIFF-AVI File Format Documentation (http://pvdtools.sourceforge.net/aviformat.txt)
- Dave Wilson's comprehensive list of FourCC codes (http://www.fourcc.org)
- OpenDML (AVI 2.0) format specification (http://www.the-labs.com/Video/odmlff2-avidef.pdf)
- AVI file format documentation (http://www.alexander-noe.com/video/documentation/avi.pdf)
- Type 1 and 2 DV AVI files (http://msdn.microsoft.com/en-us/library/windows/desktop/dd407250(v=vs.85).aspx)

Article Sources and Contributors

Technika SH-Z625 *Source*: http://en.wikipedia.org/w/index.php?title=Technika_SH-Z625 *Contributors*: Billlion, Headbomb, Martarius

Tesco *Source*: http://en.wikipedia.org/w/index.php?title=Tesco *Contributors*: (, 0 A-Dawg 0, 0404343m, 04keera, 05keera, 07keera, 08glasgow08, 08keera, 12345 wiki, 12345ar, 159753, 16@r, 1990bacon, 1e2e3e, 1st 4 Kirkby, 212 not out, 25or6to4, 28421u2232nfenfcenc, 2help, 5 albert square, 50fifty, 93gregsonl2, A Train, A bit iffy, A8UDI, ACEOREVIVED, AP1787, ARC Gritt, Aazmond, Abbiee mon, Abdullahazzam, Abdullais4u, Abfab27, Acalamari, Achmelvic, Acl499, Acroterion, Acs4b, Ad Nauseam, Adam1516, AdamBMorgan, Adambro, Adamhauner, Adamio, Adeiltrout345, Admiral Valdemar, Aegwynn, AgentCDE, Ahoerstemeier, Aidy70060, Aiken drum, Aillema, Aiman abmajid, Aitias, Aksi great, Alan Pascoe, Alansohn, Alchemist Jack, Ale jrb, Alex MacGregor, Alex43223, AlexJFox, Alexf, Aliw136, Almufasa, Alperen, Andersrask1977, AndrewSE19, AndrewWTaylor, Andy, Andycjp, Andymarczak, AngChenrui, AngeAnge22, Angela, Angr, Angusmclellan, Anna Lincoln, Ant parker, Ant0007, Antandrus, Anthony Bradbury, Anthukman06, Antony2007, Anymoon, Apau98, Aprhys, ArchStanton69, Arcturus, ArielGold, Arthur7171, Asdfasdf1231234, AtheWeatherman, Aubergene, Austrian, Avebury, Avnjay, Avoided, AxG, Axelr3, BVRT11, Baberlp, Bachrach44, BanyanTree, Barnabypage, Barrold, Barryob, Ben6993, Benabomb11, Benbread, Benchik, Bender235, Benls, Beno1000, Benwildeboer, Bettia, Bewildebeast, Bgwhite, Bhoeble, Bigasohn, Bigbluebob, Biggerbetterfastermore, Bigjimr, Bigturnip, Bihco, Bilky asko, Bjmrb, Blakkandekka, Blanchardb, Blankfrackis, Blutak11, Bob Castle, Bob schwartz, Bobblewik, Bobcat Brawler, Bobo192, Bobtkmaxfreed30, Bodnotbod, Boffert, Boleyn3, Bongwarrior, BradM, Braders17, Bradleyspencer1983, Brandsen, Brian.Burnell, Brianyoumans, BrownHairedGirl, Bubba hotep, Bucephalus, BulldozerD11, Burto88, Busy Stubber, C.byron, C2r, CALR, CBDunkerson, CGorman, CR7, Cakkum inghamb, Caknuck, CalJW, Calabe1992, Calsicol, Caltas, CalumH93, Calvin 1998, Calvin08, CambridgeBayWeather, Can't sleep, clown will eat me, CanadianLinuxUser, CanisRufus, Canterbury Tail, Canthusus, Carbonix, Careful Cowboy, Carina22, Cartrain, Caspian blue, Catapult, Catgut, Cathar11, Catrionaobrien, Causa sui, Cc0812, Cenarium, Centuriono, Chapman 07, Chase me ladies, I'm the Cavalry, Cheesy mike, Chemboro, Chester Markel, Chickenfeed9, Childzy, Chinesestar, Chirimi1906, Chocolateboy, Chodorkovskiy, Chris 42, Chris Bainbridge, Chris Wilde, Chris1193, Chrisloader, Chrisminter, Chriswiki, Chuunen Baka, Cidcn, Cimon Avaro, Claudetc, Clebb1, Clifford70, Clocker, Closedmouth, Cls14, Cm1989, Cobi, Cohesion, Colonies Chris, Cometstyles, CommonsDelinker, Coolhawks88, Cooper-42, Corvus cornix, Costesseyboy, Cpapadopoulos, Crabula, Crazytales, Crazzyquennie, Crospaul, Culnacreann, Czc06wtu, D 2 da a, D99figge, DARTH SIDIOUS 2, DMacks, Daemonic Kangaroo, Dale Arnett, Dan100, Danmush, Danno uk, Danski14, Dark Halo, DarkFalls, Darkieboy236, Darryl.matheson, Daryl 95, Dave0287, Davebelcher, Davewild, David Biddulph, DavidJonesP, Davies89, Daviessimo, Dazaclaza, Dbertman, Dcastlebeck, Dcfc92, Dchaffey, DeLarge, DeadEyeArrow, Deadflagblues, Decltype, Defiled2467, Deftera, Dell9300, Delldot, Dendodge, Denisarona, Dennisc24, DerHexer, Devcfc, Dexter000, Dgies, Dgrn1970, Diego Grez, Digifiend, Dillonlaw, Discospinster, DisillusionedBitterAndKnackered, DocendoDiscimus, Doctorwhofan63, Doktorrob, Dontknowwhereimheaded, Doochieman, Dormskirk, Dpmarshall, Dppowell, Dr h2w, Dr. Universe, Dreaded Walrus, Dreadstar, Drunkenmonkey, Dsparkes2, Dstom, Dtcdthingy, Dtcfc6991, Dunk, Durin, Dvdjturnbull, East718, Eball, Eddy.boi, Edgar181, Edison, Editor5807, Edward, Egil, El0i, Eleland, Elliot9, Emsley, Enotayokel, Eoz12, Epbr123, EphraimL, Equalizer2011, Erik1980, Erik9, Estoy Aquí, Etonian, Eurosong, Ever388, Everyking, Evilcompany, Evlekis, Exabytescom, Excirial, Explicit, Ezhiki, FC360, FCSundae, Fair Deal, Falcon8765, Farras Octara, Fatal!ty, Fatpratmatt, Favonian, Fazdabest, Fernando, Fideo, Fieldday-sunday, Filper01, Fitzebwoy, Fjlfjjl, Flewis, FlyingFlamingo, FlyingToaster, Fofr, Footyfanatic3000, Fordan, Fordy304, Fortdj33, Fortnum, Fourohfour, Francs2000, Freddyvoorhees, Frees, Fæ, GDonato, Gab86, Gaius Cornelius, Galoubet, Gammondog, Gazimoff, Gdawkins, GeoGreg, Gerbon689, Ghouston, Gimboid13, Gingerclan, Gjbloom, Glapots, Glic16, Gnowor, Gob2, Gogo Dodo, Goldom, Good Olfactory, GoodDamon, GooglesBetter, Gottago, Gr1st, Graham Jones, Greenrolo, Greenshed, Greg Grahame, GregAsche, Greswik, Grunners, Gsarwa, Gtannen, Guagd, Gunnar Larsson, Gurch, Guru Larry, Gwernol, Gwretger, H1523702, HAWC145, HJ Mitchell, Hairyhatman5, Halsteadk, HamburgerRadio, Hammodt, Haribo23, Harland1, Harry the Dirty Dog, Hathershaw, Haukurth, Hazzaahazzaa, Headbomb, Hedley Proctor, Hellorory, Henry W. Schmitt, Heron, HexaChord, HiDrNick, High venom, Himasaram, Honbicot, Hongshi, Htanna, Hu12, Huga, Huntingtower, Hut 8.5, I.am.lost, Iain Cheyne, IainUK, Ian2611, Ibshyderabad1, Icecold, Icethinking, Igorwindsor, Illictic, Imperial Monarch, Imran, IncognitoErgoSum, Inka 888, Inksmithy, Inkypaws, Inter16, Iridescent, Ismartyparty, ItsZippy, Ixfd64, J Di, J.P.Lon, J.delanoy, J45fg, JFreeman, JIP, JK23, Jack3594, JackSchmidt, Jackhonour, Jadb57, JakobSpooner, JakubMan, Jamcib, James50567, JamesAM, JamesBWatson, Jamie W, Jamse, Japanesevehicles100, Jaredbelch, Jbudgie, Jchristie7, Jdarnell010, Jeff G., Jeni, Jeremy68, Jerry teps, Jerryobject, Jfire, Jg9945, Jheald, Jhessela, Jim.Liu, Jimu16, Jjhammer92, Jkeene, Jkt, Jmb, Jmorrisseyd, Joccky A. Wilson's "shooe ties" hat, JodyB, JoeBaldwin, Joeb753, Joevictor, John, John of Reading, Johnwalton, Jonnysteel, Jooler, Jordaneas, JordieBFFtia, Josh Parris, Joshden, Joshii, Joshurtree, Joshygoodwin, Jusdafax, Juux, Jvandyke, Jza84, Jæs, Kae1is, Kakanana87, Kanyongyow, Karenjc, Katebish, Katieh5584, Kcordina, Keighleyboy, Keilana, Keiran-94, Keith D, Kevin Steinhardt, Khurram--Malik, Kiddie Techie, Kierant, Kimir, King Lopez, King of Hearts, Kingofthelemurs, Kingpin13, KirtjE, Klangboy, Klvncw, Knifespoon, KnowledgeOfSelf, Koavf, Kookas, Kpjas, Krich, Kristen Eriksen, Kristov halski, Krlply, Kuged8764, Kuhn3, Kungfuadam, Kuru, Kuzaar, L Kensington, Laagerfeld, Lajsikonik, Lame Name, LeaveSleaves, Lee Stanley, Legendary Guy, Legnaw, Lenzar, LeonWhite, Leujohn, Lifeboatsrule, Lightmouse, Ligulem, Lil crazy thing, Linkspamremover, Little Mountain 5, LittleOldMe old, Lkhn561, Longende., Lordcrusade9, LorenzoB, Lost tourist, Loudrocksurfer, LouisBenjaminCook, Louofthedead, Luk, Lukeshell2010, Luna Santin, Lynbarn, M0RHI, MBisanz, MER-C, MONGO, MRSC, Mabbb, Macduy, MadSproute, Madhero88, Magioladitis, Magog the Ogre, Majorly, Mal4mac, Malo, Mannafredo, Manors, MapsMan, Marek69, Mark83, Mark999, MarkGallagher, MarkS, Markjamesgavinprick, Marshallofteraf, Martin Jensen, Martin Peter Scott, Martpol, Matnkat, Matt.smart, Mattb210, Mattbuck, Mattgirling, Matthew Fennell, Matthew Platts, Mattmanic, Matty619, Mattyboiii, Mattyp116, Mauls, Max Naylor, Maxí, Mazca, Mc Sticks, McGeddon, McNoddy, Mchillingworth, Mdebets, Mean as custard, Meekywiki, Melissa130, Mephistophelian, Mervyn, MestUp-RiotGirl, Mgaved, Mgespeed, Mibblepedia, Michael Devore, Michaeldsuarez, MickMacNee, Mikahopalongachenkov, Mike Peel, Mild Bill Hiccup, Millisits, Mintguy, Mnbf9rca, Mondalaci, Monicatan GPEA, Moping Around, Mormegil, Morwen, Mouchoir le Souris, Mowsbury, Mr intoodeep, Mrkool123321, Ms2ger, Mtaylor848, Mufka, Mupparion, Mushroom, Mustrum, Myanw, Mynameisjesus, N1k2k8, N432138, NAHID, NCurse, NHRHS2010, Nageh, Nakon, Nancy, Nandesuka, Narom, NcSchu, Ncfcwilliam, NeilTarrant, Nengscoz416, Neothinker, Nev1, NewEnglandYankee, Nick, Nickmcfc, Nickonestop, Night Gyr, Nn123645, Nonomy, Norm, Northoltjohn, NotMuchToSay, Notinasnaid, Nouse4aname, Nsaa, Nuge, Nydas, Oatey, Obeyyourmind, Ocaasi, Oda Mari, Odysea, Ohconfucius, Oli120792, Oliekirk, Oliver Chettle, Olivier, Ont, Oo64eva, Orange Suede Sofa, Oscarthecat, Oswax, Otisjimmy1, Owen4004, Oxfordmale, Oxymoron83, PAGEMAKER, Pablo X, Pafcool, Pafcool2, Pah43, Palefire, Pan Peter, PandD Productions, Paul 012, Paul Erik, Paul MacDermott, Paul Smith 132, PaulHereNow, Pcgardner, Pcpcpc, Pearle, Pedroson, PeeJay2K3, Peridon, Persian Poet Gal, Pete147, Peteinterpol, Peter Clarke, Peter Suranyi, Peter.nyks, Petermartindill, Petrb, Petrolmaps, Pfowles, Phantomsteve, Phgao, Philip Trueman, Piano non troppo, Pigsonthewing, Pilatus, PimRijkee, Pinkadelica, Pinnecco, Plastikspork, Pne, Pointillist, Pondle, Poole14, Pork11, Portsmouthtv, Pow el er 2, Premlaldr, Pressforaction, ProcEnforce, Professor Prestwood, Pscott22, Pseudomonas, Puffin, Qst, Quantpole, QueenCake, Quiggers1P, Qwfp, RTG, Rabbayazza, RallyGames, Rangoon11, RapidAssistant, Ras, RaseaC, Ratibgreat, RattusMaximus, RaviC, Rd232, Rdc525, Reaper Eternal, Rebbeca, Rebroad, Redvers, Regancy42, Rehnn83, Reinyday, Renaissancee, Rettetast, RevelationDirect, RexNL, Rhobite, Rhollenton, Riana, Rich Farmbrough, Richard Arthur Norton (1958-), Richard Harvey, Rick030391, Rjwilmsi, Rl, Rnt20, RobertG, Rodhullandemu, Rogerd, Roketjack, Ronhjones, Rorivo, Rosestiles, Rosiestep, RossMM, RossPatterson, Rueyfgugdt, Runnermonkey, Rush786, RxS, RyanGerbil10, SGBailey, SP-KP, STopCat, Sadalmelik, Sainthalo, Sam Blacketer, Sam coles666, Sambeeni, Samcotterell, Samvanewyk, Sansonic, Santanaquintass, Sardanaphalus, Satoshi Aburatani, Satyricon uk, Scaggeh, Scgtrp, Schween, Scott5114, Scriberius, Scumbag92, Seaphoto, Secretlondon, Selbs101, Serein (renamed because of SUL), Sertrel, Shadowjams, Shadowlynk, Shariq r82, ShaunL, Shawman20, Shawnc, Signalhead, SilkTork, Simon Moss, SimonD, Simply south, Sjakkalle, Sjc07, Skarebo, Sketchmoose, Skinnyweed, Slashme, Sloman, Slough093q, Sloughlin, Slysplace, Smalkin155, Small-town hero, Smalljim, Smileyrepublic, Smjg, Smoothy, Snellios, Snigbrook, Snow storm in Eastern Asia, Snowmanradio, Snowolf, SoWhy, Sohjam20, Some jerk on the Internet, Soumyasch, Spankthecrumpet, Speed Air Man, Speedway, Spencer, Spike Wilbury, Spinner4321, SpuriousQ, Staceyizace, Starflyer, Starrycupz, StaticGull, StephenDawson, Stephenb, Steven Weston, Stevvvv4444, Stewartpalmer, Stifle, Stjarna, Storkk, Stratman07, Struway2, Suavemart, Subtractive, SunCreator, Super Ted, Svick, Synchronism, TFOWR, TGFtim, THEN WHO WAS PHONE?, Tangotango, Tankiona, TastyPoutine, Ted Ted, Tedickey, Tellyads, Tendaera, Terence, TerriersFan, Tesco Press Office, Tescohater, Tesi1700, Tgeairn, Tharnton345, Thaurisil, The Anome, The JPS, The Thing That Should Not Be, The wub, TheKMan, TheRanger, TheSnake90, Thebogusman, Thejetset1, TheoClarke, Theroadislong, Thetruth97, Thingg, Think outside the box, Thiseye, Thomas Blomberg, Thue, ThwartedEfforts, Timo74, Tjq, Tkynerd, Tnxman307, Toast, Tokailoverock, Tolkny, Tom Morris, Tom harrison, TomB123, TomGreen, TomPhil, Tomhodkin, Tomhoskin1, Tommy2010, Tommy23, Tomsk71, Tomwerner, Tonywalton, Torika, Torqueing, Tozer, Tpbradbury, Travelbird, Tregoweth, Trentondjerzi, Tresiden, Trevj, Trident13, TubularWorld, Turian, TutterMouse, Two hundred percent, Tybjvgb7uykfrd, Tyranic-moron, Udibi, UkPaolo, Ukeu, Ukexpat, Ulric1313, Ultrashaman, Uncertaincurtain, Uncle Dick, Unomi, Utcursch, VMlemon, Vanderstorm, Vandin, Vanished User 8a9b4725f8376, Vclaw, Vegaswikian, Velella, Verne Equinox, Victor Gauntlet, Visa2007, Vishnava, Voyager78906, Vrenator, WATP, WOSlinker, Wackojacko8, Wackojacko807, Wainson, Wales911, Wardy125, Watercolour12, WaveRunner85, Wayne Slam, Weedwhacker128, Weiteck, Welshleprechaun, Wereon, WhisperToMe, Whistler, Whitebox, Wiki alf, Wikidea, WillMak050389, William Avery, Wimt, Windbear Chime Guy, Wing lei, Winhunter, Wizelworthuk, Wobbletang, Woohookitty, Wrh1973, Www.continuitycorner.com, X201, XChrissehx, XDSxKibbles, Xeon0909, Xhyljen, Xogog, Xpli2000, Xsjadouk, YUL89YYZ, Ydigkhgjhtghyk, Yeanold Viskersenn, Yilmazuckez, Ylimedixie, Yomangani, Yyyooogypo, Zanimum, ZimZalaBim, Zola zazi, Zolchew, Zootm, Zulu777, Zzuuzz, לארשי הווקתח, يدماغدمحا.24, , 2584 anonymous edits

Pixel *Source*: http://en.wikipedia.org/w/index.php?title=Pixel *Contributors*: -Majestic-, 28421u2232nfenfcenc, 5 albert square, A9gould1, AGToth, Aditya, AdjustShift, Adwiii, Agateller, Agentbla, Aggsal, Ahoerstemeier, Akhilhello6, Aldie, Alex43223, Ali, Amniarix, Anclation, Andreas Kaufmann, Andres, Andrevan, AndrewWTaylor, Andrewpmk, Andries, Angus Lepper, Ap, Ariedartin, Arny, Arpingstone, Arthena, Askewchan, Astronouth7303, Atchius, Atif.t2, Auntof6, AxG, BarretBonden, Barsoomian, Basu, Bazza1971, Bearcat, Bedelato, Bender235, Betacommand, BiT, Binba, Bit, Blahma, Bobo192, Bongwarrior, Bookofjude, Bornhj, Brandt Luke Zorn, Bricevdm, Brockert, Bryan.burgers, Bubba hotep, C0nanPayne, Caltas, Can't sleep, clown will eat me, Catgut, CelticJobber, Centrx, Cgbuff, Chensiyuan, Chsimps, Closedmouth, Conversion script, Crissov, Cryptic, DARTH SIDIOUS 2, Dante Alighieri, Dave3457, Davepape, David Levy, Dbenbenn, Dcljr, Denisarona, Dennis Valeev, DerHexer, Dhatfield, Dicklyon, Discospinster, Dkstraw, Dmmaus, Dogsthatchasecars, Donarreiskoffer, Dontknowhow, Dosnerd90, Doug Bell, Dreg743, Drivenapart, ESkog, Ed g2s, Edward, ElGordo, Electron9, Eliz81, Elkman, Ellmist, Ellywa, Enos Shenk, Epbr123, Erik9, Ethaniscool1, FCYTravis, Falcon8765, Fastfission, Ferengi, Feydey, FiP, Firsfron, Fishnet37222, Fleminra, Flewis, ForestJay, Frecklefoot, Fritz Jörn, Fuhghettaboutit, Funandtrvl, Fusionmix, Gaius Cornelius, Galoubet, Gbrugman, Georgec, Ghakko, Ghewgill, Giftlite, Gits (Neo), Gogo Dodo, Goingstuckey, Graham87, Guppy, Gurubrahma, Haakon, Hadal, Haham hanuka, Hamish2k, Hamtechperson, Handy helper, Haseo9999, Hephaestos, Heracles31, Heron, Hfastedge, Hkpawn, Imroy, InverseHypercube, Ioapetraka, Ixfd64, J.delanoy, JIP, Janke, JayJayKay, Jeff G., Jeffrey O. Gustafson, Jesse Viviano, Jfingers88, Jim.henderson, Jk2q3jrklse, JuanFox, Juliancolton, Karada, Kbdank71, Keenan Pepper, Keilana, Kieff, Kim Bruning, Kim0kim0, Kirill Lokshin, Kjkolb, Kjlewis, Kpjas, Kprateek88, Kwaku, LOL, LarryGilbert, Ld100, Liftarn, Locador, Localdara, Lordofthe9, LuckyIMstar, Lumpbucket, Lysdexia, M.nelson, MARKYSPARKY19, MER-C, Malo, Marasmusine, Mark83, Mattisse, Maxim, Mdd, Merlinme, Merphant, Miaow Miaow, Michael Hardy, Midway, Mikeipedia, Miraceti, Mlaine, Moloch09, Morken, Moxfyre, MrMambo, Mwhiteguy, NathanHagen, NawlinWiki, Necronomicronistic, Nimbusania, Nmacu, Nothingofwater, Ohnoitsjamie, Omegatron, OnBeyondZebrax, OnePt618, Orderud, Oyd11, Patrick, Pearle, Pengo, Pete Peterson, Pforret, Philippe,

Pilotguy, Pinball22, Pinethicket, Pinkadelica, Pizza Puzzle, Planemad, Poeloq, Preslethe, Prewitt81, Pshent, Puchiko, Punch, Drunk, Love, Purpledramallama, Pursin1, Quota, Qxz, Qz, Ran4, Rebornsoldier, RedHillian, Reedbeta, Registreernu, RexNL, Reyk, Ricky81682, Rob215, Robert Buzink, Rocket000, Rory096, Roux, Rrburke, Ruslik0, SJP, Sabisteven, Sam Korn, Samuell, Samus3456, Santrobi, Sceptre, Scott McNay, ScriptedGhost, Seabhcan, Seb, Sebleblanc, Secfan, Sergeyy, Shawn in Montreal, Shirik, Silver Edge, Simeon, SimonP, SimonPowell, Sionus, Sjanuary, Slokunshialgo, Sluzzelin, Smalljim, SmartAVI, Snowolf, SouthernMyst, Speck-Made, Srleffler, Staffwaterboy, Sum0, Svick, TAKASUGI Shinji, Tarquin, The wub, TheMindsEye, Thecheesykid, Themepark, This, that and the other, Thurs Day, Thw1309, TimMagic, Timothy1808, Tiptoety, Tombomp, Tony1, Traveler100, Trekphiler, Tuspm, Urdna, Utcursch, VMS Mosaic, Verpies, Violetriga, Vladsinger, Vrenator, Wapcaplet, Wavelength, Wernher, West.andrew.g, Wik, Wiki alf, Wikid77, Wikidudeman, Wtshymanski, Xenoranger, Yekrats, Yudiweb, ZacBowling, Zanimum, Zdog18, ZeroOne, Zigzagfes, Zlobny, Zundark, Михајло Анђелковић, 621 anonymous edits

Digital camera *Source*: http://en.wikipedia.org/w/index.php?title=Digital_camera *Contributors*: -Majestic-, 4wd, 5 albert square, 7castle, 842U, A little insignificant, Aaron Brenneman, Acroterion, Adambro, AdeMiami, Admbws, Adoniscik, Aecis, Aflm, Ahmed Elnagar, Ahoerstemeier, Alan Liefting, Alanbrowne, Alansohn, AlbertCahalan, Aldebaran66, Alecv, Alethe, Alexius08, AlistairMcMillan, Almazi, AmarChandra, Anastrophe, Andrewpmk, Aniksumon, Ankeypan, Anonymous editor, Anthony22, Antilived, Arathald, Arjun01, Arthur2045, Ashear, Atakdoug, Atlant, Atmozaps, Atreyu rockes wohoo, AwamerT, Badgernet, Barek, Barry Pearson, Barry.hendy, Baylink, Beetstra, Betbest1, Bettia, Bijee, Binksternet, Birdmessenger, Bissinger, BlaiseFEgan, Bobo192, Bongwarrior, Brokenspirits, Brownwicker, Brusegadi, C0N6R355, CA387, CLW, Calbookaddict, Caltas, Cambridgeincolour, Can't sleep, clown will eat me, Capricorn42, Cburnett, Cdmarcus, Celestianpower, Ceyockey, Chalkstar2188, CharlesB02, CharlesC, Charlottiepants, Chealer, Chris the speller, Chrisjj3, Christopher-brent, Christopherlin, Ckatz, ClarkMills, Cogiati, Cohesion, Commander Keane, CommonsDelinker, Coreyfinewood, Cryogenius, Csichris, Cyril42e, DShantz, DVdm, Daemonaka, Dale Arnett, Dancter, DanielCD, Danntm, Dantheman531, Daqu, Darx9url, Darz Mol, DaveLoneRanger, Dcb1995, Dck7777, DeadEyeArrow, Denisarona, Deor, Desherinka, Dicklyon, Dimosvki, Dipics, Discospinster, DivineAlpha, Divok, Djmckee1, Dmol, DocWatson42, Doh5678, Donarreiskoffer, Dotancohen, Dragon 280, Dtcdthingy, DynamoDegsy, Dysepsion, Ecallison2010, Ed g2s, Edgar181, Edward Z. Yang, Efficacy, Egghead06, Egil, Ek13, Elvenscout742, Emote, Emoticon, EngineerScotty, Ente75, Epbr123, Epson291, Epylar, Ericg33, Erotml, Escape Orbit, EstebanF, EurekaLott, Eventhiel, EverettColdwell, Excirial, ExpertsComment, Falconssb1, Fauxcouture, Favonian, Fdp, Febcom, Feinoha, Fiftysixmoha, Fir0002, Firstorm, Flasher, Foobar, Fourthords, Ftpaddict, Fultron89, GB fan, Gabbe, Gafex, Gail, Galwhaa, Gazimoff, George100, GermanX, Ghosthost, Giftlite, Gilliam, Gillyroze, Girolamo Savonarola, Glen7565, Glipgh, Globalbattery, GraemeL, GregorB, Gregorof, Greif, Gsarwa, Gubak, Gurch, Gurubrahma, Gz33, Hadal, Hamtechperson, Haroon85, Hdt83, Hellomaster5, Henrik, HexaChord, Hosse, Husky, Hut 8.5, Hywel, IGeMiNix, Iamjforlife, Idleguy, Imroy, Indiealtphreak, Indon, Iridescent, Iruka, Isaac Sanolnacov, J Milburn, J.delanoy, JForget, JNW, James11899, Jan von Erpecom, Jannev, Jasonpage, Jawsper, Jeff dean, Jeffrey Mall, Jesster79, Jim.henderson, JimVC3, JoaoRicardo, Johnchapple, Johnteslade, JonHarder, Jonathan O'Donnell, Josh Parris, Joshyfrancis, Jpgordon, Jpmaurya, Julesd, Juliancolton, Jumbuck, K-car, KC109, KGasso, Kameraad Pjotr, Kanabekobaton, Katieshaw121, Kazım, Kbh3rd, Kcordina, Keilana, Kelvie, Kevin B12, Kghose, Kid genious, Kintetsubuffalo, Kinu, Kjkolb, Kozuch, Krasimir, KrisFujiZ, Kukini, Kulaa, Kuru, Kx1186, Kyoko, Kyorosuke, KyroII, L Kensington, Latka, LeaveSleaves, Leeharasyn, Levineps, LibLord, Lightmouse, Little Mountain 5, LittleOldMe, Loganberry, Lord Darth Vader, Luckas Blade, Lupo, MER-C, MJSkia1, MK8, MONGO, MPerel, Mac, Mahaboyd, Mahlum, Mani1, Manop, Mansichoudhary.mansirana, Marc Lacoste, Marcelo1229, Marek69, Mariocki, MarkSweep, Martarius, Martinmaniac, Masterpjz9, Materialscientist, Matey, MathewTownsend, Matt B., Matt Deres, Matt0401, Matt57, MattJons, Matthardingu, Maxamegalon2000, Mbell, Mdrejhon, MediaMangler, Mentifisto, Message From Xenu, Metageek, Mhoenig, Midnightcomm, Mikael Häggström, Mikeo, MikesPlant, Mild Bill Hiccup, Mindgames11, Mindmatrix, Misou, Misternuvistor, Mlearning, Mohylek, Molerat, Monkeyman, Moreati, Motorrad-67, Mouseover, Mr Bartels, MrBurns, Mrkastorano, Mrsardo, Musiphil, Mwanner, Mxn, Mysidia, Mywebenterprises, NHRHS2010, NaBUru38, Nakon, Navstar, NawlinWiki, Nectarflowed, NellieBly, Neo-Jay, Neutrality, Newone, Nexus Goof, NickW557, NigelR, NightThree, Nixdorf, Nkthen, Nopetro, Noq, Norm mit, Northamerica1000, Notheruser, Nslimak, Nurg, Nuttycoconut, O'Dea, Oatmeal batman, Obesesumo, Ohnoitsjamie, Oicumayberight, Oli Filth, Ooza, OrgasGirl, Oxymoron83, Ozzieboy, Patriarch, Patstuart, Paul August, PaulGarner, Pavarolar, PeaceAnywhere, Peregrine981, Philip Trueman, PhilipO, Phillipsacp, Piast93, Pilatus, Pinethicket, PlanD, Pol098, Pshent, Pyfan, Quintote, Qxz, R'n'B, R69S, Raptordrew, Raviaka Ruslan, Raysonho, Redlock, RenniePet, RexNL, Rhobite, Rich257, Richardcavell, RioRico, Rjwilmsi, Rl, Rmosler2100, Robert yu@sony.ca, Rogerd, Ron Johnson, Rror, Rtcpenguin, Ryan Albrey, Rynsaha, SDuuring, STHayden, Sagar erande, Sam Korn, Samfoster25, Sandstein, Santrobi, Saros136, Sbn1984, Schneelocke, Scriberius, Sdgjake, SeanMack, Seano1, Seb26, Seeklinks, Seidenstud, Shadowjams, Shadypalm88, Shartaj1985, Shotgunlee, Siddharthmukund, Sietse Snel, Silver Streak, SimonP, Skaterjo, Sketchmoose, SkyWalker, Slicky, Smartfunda, SoWhy, Soler97, Sophie means wisdom, SpeedyGonsales, Spitfire, Splintax, Stefan, Stefano89, Stephenb, Stevertigo, Stuartyeates, Stujoe, Stwalkerster, Stybn, Sushi fish, T.otoole67, TaintedMustard, Tangotango, Tassedethe, Taxman, Technobadger, Tecoutlet, The High Fin Sperm Whale, The Letter J, The Rambling Man, The Thing That Should Not Be, TheMindsEye, Thegerm, Thegreenj, Themfromspace, Theosch, Thewikimasterforsure, Think outside the box, Thiseye, Thisisborin9, Thrapper, Tide rolls, Tillman, Tim Starling, ToNToNi, TobiWanKenobi, Tommy2010, Tommylommykins, Tonyogun, Traxman, Tubesship, TweetyPez, Ugen64, Ujjwalb13, Uncle Dick, Upupp, Vague Rant, Valley2city, VasilievVV, Velella, Versus22, Victor cam, VolatileChemical, Voyageur, Waggers, Walton One, Wavelength, Wcosshall, Weast, Weetoddid, Werratal, West.andrew.g, Wifocals, Wiki alf, WikiBone, WikiRob12, Wikigi, William Avery, Wimt, Wispanow, Wknight94, Wolfkeeper, Woohookitty, Wrs1864, Wtshymanski, WuWei, Wuffyz, Wwoods, XJamRastafire, YUL89YYZ, Yair rand, Yamamoto Ichiro, Yamla, Yoganate79, Yuckfoo, Z.E.R.O., Zebbadiah, Zhou Yu, Zigger, Ztyler90, Zundark, Zzuuzz, 1301 anonymous edits

AA battery *Source*: http://en.wikipedia.org/w/index.php?title=AA_battery *Contributors*: 041744, 16@r, 32X, Abce2, Abdull, Adambro, Alexforcefive, AlistairMcMillan, Altaphon, Amalas, Andrei901, Angrycrustacean, Anon user, Anonymous anonymous, Arm, Ash108, Avicennasis, Avochelm, Avono, BD2412, Bbrownp, Beland, BirdValiant, Bongwarrior, Bsoo, CDN99, Cat Parade, ChaChaFut, Cheeesemonger, Coasterlover1994, Colden46, Cphi, D, David Gerard, Deich, Deli nk, DocRuby, Długosz, Edison, Entheta, Epbr123, Eric-Wester, ErisDiscord, Eternalschala, Eugbug, FileTrekker, Flameviper in Exile, Gaius Cornelius, Garion96, Gbfierro, Gene Nygaard, Glenn, Greatersam, Gromlakh, Grutness, Gwernol, Harry, Harryboyles, IHRH209, Inzy, Isidore, ItsZippy, Iuhkjhk87y678, J04n, Janke, Janus1339, Jared, Jer10 95, Jim.henderson, John Smith 104668, Jp314159, Jsp3970, Koavf, Kri, Kwamikagami, L Kensington, L1A1 FAL, LOL, Lead holder, Lightmouse, Lmatt, MLRoach, Mac, Macy, Manishearth, Martarius, Mattbayly, Max Schwarz, Meanmeancoffeebean, Michaelbluejay, Moa3333, Moxfyre, Musicnotes8, NTK, NewEnglandYankee, Newmatt3, Nicob1984, Nintendude, Oldlaptop321, Palica, Pearle, Pengo, Peter Horn, PeterJeremy, Photoguy439, Pierisb, Praetorian42, Prari, ProhibitOnions, Quatermass, Quickmythril, Quickos, RNLion, Ralmin, RenOfHeavens, Rich Farmbrough, Rifleman 82, Rintrah, Rjwilmsi, Robert Brockway, Rod57, SSDPenguin, Saharabeara, Samnesral, Sbmeirow, Sbn1984, Segv11, Shablog, Siddhant, Silverxxx, Sligocki, Slippyd, SoWhy, Sole Flounder, Sonia, Steam5, Stubblyhead, Supergloom, The Thing That Should Not Be, Thefatmanwins, Theo10011, Thumperward, Tiddly Tom, Tri400, VVikipedia343, Vectro, VgerNeedsTheInfo, Voorlandt, Wayne Hardman, Wtshymanski, Xdamr, Yamla, 216 anonymous edits

Zoom lens *Source*: http://en.wikipedia.org/w/index.php?title=Zoom_lens *Contributors*: -Majestic-, A876, Adorama, Akulkis, Angela, Anoneditor, Arnero, AxelBoldt, Benitoisbackagain, Betacommand, Borgx, Cburnett, Church of emacs, Coneslayer, Conscious, Desherinka, DocWatson42, DrBob, Egil, Ericd, Erkan Umut, Fg2, Filu, ForrestCroce, Fountains of Bryn Mawr, Fourohfour, Fourthords, Francis Flinch, Fsgs, GregorB, Grendelkhan, H, Hooperbloob, Hugh2414, Imroy, Iulius, Jeffmedkeff, Jnavas, Jncraton, Joergen, JordoCo, Jwpurple, Keecheril, Kokoo, Kuru, Leemeng, Lenehey, Loganberry, Lynophi, Macaddct1984, Marc Lacoste, MarkSweep, Michael Hardy, Mindmatrix, Moxfyre, Muijz, Munford, NFH, NameIsRon, Nurg, Peter S., Petri Krohn, Pol098, RJASE1, Rama, RedAndr, Rror, SDC, Sam Hocevar, Saperaud, Scriberius, Se16teddy, Seano1, Shotgunlee, Sir Link, Srleffler, Stybn, Tamasflex, That Guy, From That Show!, Thermofan, Thumperward, Timrb, Trabelsiismail, Tubesship, Van der Hoorn, Warmanddry, West.andrew.g, Woohookitty, 70 anonymous edits

Focal length *Source*: http://en.wikipedia.org/w/index.php?title=Focal_length *Contributors*: Aaron Brenneman, AdultSwim, Ali@gwc.org.uk, Anarchitect, AndyKali, Arctic Night, Autoerrant, Bandit1125, Beland, BenFrantzDale, Betacommand, BoP, Bobo192, Borgx, Cambridgeincolour, Canadian-Bacon, Cheeni, Chitrapa, Cmglee, ConCompS, Cosmic Latte, Denelson83, Diberri, Dicklyon, DrBob, Ehn, Epbr123, Everything Else Is Taken, Fudoreaper, Galar71, Gbruin, Glenn, Gogo Dodo, Googoohead, Hakan Kayı, Hankwang, Headbomb, Henrikb4, Hooperbloob, Hugh2414, Icairns, JFreeman, Jeremy Butler, Jujutacular, KYN, KymFarnik, Liaocyed, LieAfterLie, Lisawaa, MarcGushwa, Martarius, Maxis ftw, McGeddon, Meggar, Milkmandan, Mindmatrix, Moemin05, Moxfyre, MrFloatingIP, Nbarth, Neko-chan, Nigelj, Oli Filth, OrionKnight, Pflatau, PierreAbbat, Possum, Quistnix, RainbowOfLight, Redjar, Rl, Rooh23, SharkD, Srleffler, Swales, Tamasflex, The High Fin Sperm Whale, The wub, Thebassman, Tiddly Tom, Venny85, William Graham, Wricardoh, Yomamas, Yurik, 139 anonymous edits

USB mass-storage device class *Source*: http://en.wikipedia.org/w/index.php?title=USB_mass-storage_device_class *Contributors*: 892christie, 999999user, Acdx, Alecv, AlistairMcMillan, Angelic Wraith, Bk0, Boffy b, Borehamt, Chameleon, Chuckie88, ClariT, Claunia, Coolbho3000, Craigy144, Csabo, CyberSkull, DD2K, Damian Yerrick, David Gerard, Dcorzine, Dolda2000, Dreamwalker666, EdgeOfEpsilon, Eternalschala, Etienne.navarro, Excirial, Fabulatech, Finnce, Gaborgasko, Garuchi, Georg Potthast, Gg666, Giraffedata, Gnuish, GraemeL, Grendelkhan, Groyolo, Haakon, Hello71, IMSoP, Ingvarr, Ishdarian, JDean78, JKD, JLaTondre, Jamie Lokier, Jerryobject, Jim.henderson, John Fader, JonHarder, Jw21, Kainaw, Kf4bdy, Khfan93, Laptopdude, Lee Carre, LilHelpa, Lordandmaker, Lradrama, Mac, Mako098765, Mark5677, Marko75, Martarius, Mentifisto, Mild Bill Hiccup, Minime72706, Mithent, Moxfyre, Music Sorter, Nasa-verve, Nixdorf, Nono64, Nuwewsco, Ohnoitsjamie, OlavN, Omegatron, P23y1, PTSE, Pbeesley1989, Phgao, Pipatron, Pmsyyz, Pne, Polluks, Ponydepression, Radiojon, Raffaele Megabyte, Ralphbk, Rink, Rjwilmsi, Rklawton, Ronbarak, Sai.dumpu, Scgtrp, SchmuckyTheCat, Seven of Nine, Shawn K. Quinn, Silly rabbit, Sky Attacker, Slowking Man, SpuriousQ, SufferWell1396, The Mad Bomber, Thumperward, Thunderbrand, Towel401, Trasz, Traut, Twilsonb, ValeOfAldur, William Graham, Willmcgree, Windchaser, Wrs1864, XMog, Xorxos, Xpclient, Zwart, 206 anonymous edits

USB *Source*: http://en.wikipedia.org/w/index.php?title=USB *Contributors*: 0612, 132qwerty, 1wolfblake, 223fms, 4176shelton, 5 albert square, A Train, A0183305, A5b, AJ6J, AMK1211, ANTMAN, AaronLLF, Aaronbrick, Aaweisen, Abce2, Abdul raja, Abune, Access Denied, Accounting4Taste, Accurizer, Acela Express, Achraf52, AdRock, Adambro, Addps4cat, AdeMiami, Adot, AdunaicLayman, Advantecheautomation, Aeons, Agateller, Agilent.Showard, Agitate, Ahivarn, Ahoerstemeier, Ahruman, Aido2002, Aidoor, Airplaneman, Airsplit, Ajfweb, Ajitesh Madai, Ajpvalente, Akasnakeyes, Alansohn, Aldie, Aleksandrit, Alex mayorga, Alexander.stohr, Alexei123, Algae, Algocu, Ali@gwc.org.uk, AlistairMcMillan, Allesbehalve, Alpha 4615, Alphathon, Altermike, Alvin-cs, Alyssa3467, Amdma2003, AmiAyalon1969, AnOddName, Andrew sh, Andrew19881123, AndrewLeeson, Andreweaton, Andrewpmk, Andy Marchbanks, Andyzweb, Angela, Anibius, Anonymi, Ant75, Antandrus, Antialias, Apoc2400, AppOnKey, Archer3, Arichnad, Aries21erika, Arkrishna, Arlie davis, Arm, Armando, ArnoldReinhold, Aronzak, Arthur Rubin, Asafoot, Asbjornu, Asetwofifty, Ashenai, Ashleypurdy, Asigler, Asim mahakul, Asim18, Asparagus, Atalsandip, Atamido, Atannen, Atenor1932, Atlant, Attilios, Audrius u, Austinv311, Avalyn, AxelBoldt, Axlq, Azhyd, Azuosdativa, B. van der Wee, BCG999, BZRatfink, Baoap, Baricom, Bcrscahh198987, BdON003, Bdesham, Beao, Becksguy, Beland, Belvdme, Bender2k14, Benhoyt, Benignbala, Benny45boy, BesigedB, Bestalex, Bgkwtnyqhzor, Big gun, Bitchbastardd, BjKa, Black Walnut, BlackWolf, Blacklife85, Blu3tooth, Blugill, Bo98, Bobbb53, Bobblewik, Bobo192, Bobrayner, Boijunk, Boing! said Zebedee, BorgQueen, Bratch, Brenbren92, Brewthatistrue, BrianRecchia, Brianpeiris, Brianski, BrickMcLargeHuge, Brighterorange, Brim, Brisvegas, BrockF5, Brouhaha, Brownsteve, Bruns, Bryan Derksen, Brycen, Bssasidhar, Btornado, Bubeck, Bullzeye, Bungle, Burnte, Burrin.p, Buxtehude, Bytencoder, Bz2, C0nanPayne, CALR, CAkira, CCFreak2K, CINCABF, CWenger, Cabe6403, Cactus.man, Cadre, Caiaffa, Calabraxthis, Calltech, Cambrant, CambridgeBayWeather, Can't sleep, clown will eat me, CannonR,

Capbat, Capricorn42, CaptainClawz, CaptainVideoJW, Carmichael95, Casper2k3, Caspertheghost, Caulde, Ccradio, CecilWard, Ceros, CesarB, Cgumas, Charles Gaudette, Chealer, Chiefcoolbreeze, Chimin 07, Chipp C, Chirry2011, Chovain, Chowbok, Chridd, Chris Chittleborough, Chris G, Chris53516, ChrisFAF, Christopher Parham, Chronulator, Chrumps, Chrylis, Chuunen Baka, Ciaranjns, Cj tzm, Clam0p, Classical geographer, Cleared as filed, ClickRick, Closedmouth, Closenplay, Cmcqueen1975, Cmgross, Cobaltbluetony, Codemsan, Colenso, Colfer2, Colinkgl, Colonies Chris, Cometstyles, CommonsDelinker, Compellingelegance, Compilation finished successfully, Coneslayer, Conrad.Irwin, Conversion script, CoolFox, Coolstr24, Cootiequits, Coreywalters06, Coroboy, Corrosive.element, Corvus cornix, CountFlux, Courcelles, Crazy Fox, Crispmuncher, Crissov, Crywalt, Ctachme, Ctjf83, Cwalger, Cwolfsheep, Cybercobra, Cyferz, D zone, D0762, DKqwerty, DMahalko, DSRH, Da Vynci, Daft Creftsman, DaisyChainer, DaleDe, Dalziel 86, Damian Yerrick, Dan100, Daniel Pritchard, Dank1993, Danski14, DarkFalls, Darkdawn75, Darkknight512, Darkride, DarthShrine, Dave laird, Dave6, David Biddulph, DavidCary, DavidFarmbrough, DavidMarsh, Davidcx, Davidfstr, Davidgordon101, Dcook32p, De728631, Dead3y3, DeadEyeArrow, Deh, Dekisugi, Deletros, Den fjättrade ankan, DenisBlanchette, Depakote, Derek Ross, DerekMorr, Derepi, Devin6687, Diamondland, Diderot's dreams, Dirtyfrank10, Dispenser, Dittaeva, Djbyrne17, Djcapelis, Djmckee1, Dlother, DmitTrix, DmitryKo, Dnas, Doc Magnus, DocWatson42, Dolda2000, Don-vip, DopefishJustin, Doradus, Dori, Dougher, DragonHawk, Drdomestod, Drewdc90, Ds13, Dsgreat3, Duckbill, Dvelez1985, Dysprosia, E.boyer7, E090, EagleOne, Earle Martin, Ebe123, Ed Brey, Ed Poor, Ed g2s, EdLegend, Edgar181, Editore99, Edokter, Ee79, Efitu, Egil, Egmontaz, Ehn, Eighthave, Electron9, ElementFire, Elpincha, Elsendero, Elvey, Emerson7, Energyequation, Engineerism, Enochlau, Enquire, Enviroboy, Epbr123, EpiVictor, Epson291, Equendil, Eric Wester, Espertus, Eurleif, Eus Kevin, Evan-Amos, Evert Mouw, Everyking, Evice, Evil saltine, Excirial, Eye of God, Eyreland, Ezhuttukari, F l a n k e r, FT2, Fabulatech, Facts707, Fahidka, Failofbeaner, False vacuum, Falsifian, Favonian, Feedmecereal, FelisLeo, Ffgamera, Fibonacci, Fieldday-sunday, Filceolaire, Filelakeshoe, Filzstift, Finnegar, Fir0002, Fireaxe888, Fishnet37222, Flazh, Flightsoffancy, Floorwalker, FlyingToaster, Fox, Frammy7, Frap, Fredrik, Fromageestciel, Fudoreaper, Fumo7887, Func, Fustigate314159, Fuzzie, Fvw, G2M, GDonato, Gabbe, Gail, Gaius Cornelius, Gardar Rurak, Gauravsangwan, Geekstuff, Geniac, Georgy90, Gh5046, Ghartwig, Ghermann3, Giftlite, Gilliam, Ginsengbomb, GioCM, Giraffedata, Givemeornot, Glenn, Gogo Dodo, Golbez, GoldKanga, Good Intentions, Goodone121, Gordeonbleu, Gosale, Gpearson2, Graglin, Grand Am, GrandMoffVixen, Grapetonix, Gregben, GregorB, Gregsea, Grendelkhan, Groink, Groogle, Gryffon5147, Gschizas, Gssq, Guaka, Gutza, Guyjohnston, Gyll, H2g2bob, H3llbringer, Ha runner, Hankwang, Hans Dunkelberg, Hargrimm, Harvester, Haseo9999, Hayabusa future, Hcaaman, Hcs, Hdante, Hduckman, Hellgi, Heron, Hgrosser, Hideyuki, Hlandro77, Hoemaco, Hohohob, Homerjay, Honeycake, Hoof Hearted, Hooperbloob, Hopp, HorsePunchKid, Hughcharlesparker, HumanJHawkins, Hundehalter, Husky, Hydrargyrum, Hyins, I am a true NOS person, I dream of horses, IAMBATMANDEALWITHIT, INVERTED, ISquishy, Ian Pitchford, Ian01, IanGM, Iandiver, Icairns, IceSlicer, IcedNut, Icydid, Ilia Kr., Iloveandersoncooper, Immunize, Imotor, Impasse, Imperator3733, Imroy, Inclusivedisjunction, Intelligentsium, InternetMeme, Intersofia, Intgr, IntrigueBlue, Iridescent, Irishguy, Isaac, ItsZippy, Iuhkjhk87y678, Ixfd64, J.delanoy, J.smith, JCLAWSON, JDX, JHP, JLaTondre, JLuna103, JR98664, JTN, JaGa, Jabberwoch, Jacob Poon, Jake Wartenberg, James Kemp, Janto, Jargoness, Jason One, Jasper Deng, Jaufrec, Jay-the-mad'n, JayC, Jcarroll, Jdthood, Jdwinx, Je007, Jean-Baptiste Catté, Jeffq, Jeffro77, Jeffthejiff, Jeltz, JeremyA, Jerryobject, Jesse Viviano, Jesslovesmeganandpie, Jesster79, Jezmck, Jhsounds, Jidanni, Jim A H, Jim-Bob Harris, Jim.henderson, Jimgawn, Jimmi Hugh, Jimp, Jjjjoooooeeee123, Jkl4201, Jmayorga5, Jmrowland, Jmundo, Jnavas, Jncraton, Jnsears, JoanneB, Joaopaulo1511, JoeOnSunset, Joeinwap, Joel D. Reid, Johammer, John Fader, John Reaves, JohnCD, JohnSawyer, Johnteslade, Jon125, JonHarder, JonSangster, Jonathan Grynspan, Jonathunder, Jonel, Jonverve, Jordan Brown, JosephCampisi, Josh Parris, Joshua Scott, Jossi, Jpgordon, Jruderman, Jrvz, JtMinahan, Julesd, Julienrl, JustinRossi, Juux, Jwidjaja, Jwinius, Jwoodger, KJRehberg, KUsam, KVeil, Kajaco2, Kalan, Kaldosh, Karn, Katanada, Katyare, Kazvorpal, Kbh3rd, Kbolino, Kbrose, Keilana, Kelleheretic, KelleyCook, Kemiv, Kendal Ozzel, Kenny sh, Kenyon, KerryVeenstra, Kghose, Khalid hassani, Kia.Rahimi, Kilo-Lima, King of Hearts, Kiteinthewind, Kittens-Pedro, Kjkolb, Kjohna, Kniesten, KnowledgeOfSelf, Knuckleskin, Kocio, Kooo, Koopa turtle, Kostmo, Kozuch, Krallja, Krash, Krishvanth, Kristof vt, Ktr101, Kudret abi, Kuru, Kvng, L Kensington, LN2, La Pianista, Lab16, Lada103, Laefer, Lambtron, Landroo, Lankiveil, LarryLACa, Lawpjc, Leandrod, Letdorf, Lhopitalified, Lightmouse, Linas, Ling Kah Jai, Link83, Lipatden, Lithos12, Lithpiperpilot, Little Mountain 5, Livebird, LivingShadow, Lockoom, LodeRunner, Logan, Lonaowna, LorenzoB, Lotje, LouScheffer, Lovely Chris, Lowellian, Lowenddan, Lucasreddinger, Luna Santin, Lupo, M1ss1ontomars2k4, MCG, MER-C, MISTYFAN4EVER8887, MNAdam, MSR93, MaGioZal, Maaf, Mabdul, Mac, Mad with power, Madison Alex, Manassehkatz, Mange01, Manishearth, Manuactive, Manuelt15, Margin1522, Mark Yen, Markhoney, Markstuart44, Marstronix, Martarius, Maschneider, Materialscientist, Matt Crypto, Matta33178, Mauls, Maury Markowitz, MaxHund, Maximus Rex, Maxis ftw, Maxvip, Maxwellversion2, Maxí, Mayhemm, Mazin07, McGeddon, McSly, Mcarling, Mcorazao, Meaningful Username, Mendaliv, Mentifisto, Merlinsorca, Mewtu, Mgdunn, Michael Hardy, Michaelkourlas, MidMadWiki, Midlandstoday, MightyWarrior, Mike.lifeguard, Mike1024, Mikeblas, Mikus, Mild Bill Hiccup, MilfordBoy991, Miquonranger03, Mirddes, Misocroft, Mj fan1995, Mjpieters, Mkdw, Mkouklis, Mlewis000, Mnw2000, Mobius, ModsRule, Modster, Mojo-chan, Morcheeba, Moreati, Mortense, Moxfyre, Mpa, Mr Minchin, Mr.Z-man, MrBurns, MrDolomite, MrFish, MrOllie, MrSomeone, Mrappleton, Mrschimpf, Mrtangent, Mschlindwein, Muad, Mugunth Kumar, Muhandes, Mulad, Mun206, Munge, Mushroom, Music Sorter, Mvjs, Mwarren us, Mwilso24, Mxjose, My man Friday, Mysekurity, Mysidia, N2e, N5iln, NHRHS2010, NZR, Nageh, Nahaj, Nanshu, Naohiro19, Narge, Nasukaren, NawlinWiki, Nchalada, Neil916, Neilm, NellieBly, Netkinetic, Niceter, Nick, NickVeys, Nikkibella21, Nikpapag, Nil Einne, Nishantjr, Nixdorf, Nk, Nlaporte, Nnetala, Noahspurrier, Nopetro, Norm, Nrbelex, Nsaa, Nubi78, Nullaman, Nurg, Nw15062, O18, Odam, Odatus, Oehoeboeroe, Ohnoitsjamie, OlEnglish, Old Moonraker, Olekrst, OllieFury, OllyH, Oman9978, Omegared23, Omegatron, Omicronpersei8, Oneiros, Oni Ookami Alfador, OpenToppedBus, Osarius, Osram, Ottawa4ever, Outlyer, OverlordQ, Owengibbins, Oxymoron83, PRRfan, Pabouk, Padillah, Paradoctor, Paranoid, Parmastew, Patriarch, PaulColby, PaulMcCulloh, Pavel.nps, Pbacina, Pchov, Pcuser42, Peak, Pedant17, Pedro, Pembers, Perardi, Perfectblue97, Perryizgr8, Persian Poet Gal, Peruvianllama, PeterGrecian, Petiatil, Peyre, Pgan002, Pgiii, Phantasee, Phantomsteve, Phatom87, Phgao, PhilKnight, Philip Trueman, Phip, Phobie, Phoenix314, Photographerguy, PiMaster3, PiaH, Pie4all88, Pietrow, Pinkadelica, Pinkcious, Pip2andahalf, Pivotto, PizzaMan, Pjrm, PlatinumX, PlayStation 69, Plugwash, PluniAlmoni, Pmc, Pne, Pointillist, Pol098, Polluks, Poppafuze, Prari, Praveentech, Preslethe, ProhibitOnions, Puchiko, Pugetbill, Pugglewuggle, Pushpinder86, Qasimnb, Qk, Quadell, Quentin Jones, Quinxorin, Quizzicus, Qxz, R!SC, R'n'B, R2D2 C3PO R2D2, RAMChYLD, RJHall, RadioactiveKiller1, Radiojon, Ragzouken, Ramu50, Randomperson666111, Raptor007, RasputinAXP, Ravenperch, Ravensfan5252, Raymond Hill, Raysonho, Rbellika, Rbrittner, Rchandra, Rcingham, Rdnetto, Reach Out to the Truth, Realist2, Rearden9, Rebroad, Red, RedWolf, Redsully, Reeceyyyy15, RegentsPark, Reisio, Remember the dot, Remove indian propaganda, Res2216firestar, Retodon8, Revera, Revrant, RexNL, Rexrodo, Rfc1394, Rfl, Rhindle, Rhobite, Rich Farmbrough, Richardcavell, Richardpitt, Ridge Runner, Rikonate, Ringbang, Rinix, Rip969, RitKill, Rjwilmsi, Rmhand, Rmsuperstar99, Rob Cranfill, Robert Loring, RobertG, Rocastelo, RockMaestro, Rodneyorpheus, Rohasnagpal, Romanm, Ronark, Ronhjones, RoninRVP, Rossheth, Rpkrawczyk, Rprpr, RufusThorne, Rupert Clayton, RussNelson, Russella, Rwestafer, Rwwww, Ryper, S. Neuman, S.K., SLi, SMC, Sadalmelik, Sadharan, SafariSunD, Saimhe, Salamurai, Salavat, Salvio giuliano, Sam Hocevar, Samarqandi, Samuel Grant, Sango123, Sanspeur, Sasuke Sarutobi, Saulo Paiva, Savant13, Sawyeriii, Sbmeirow, SchmuckyTheCat, Scollk, Scope creep, Scuac, Sdsds, Seanmcd27, Seidenstud, Semicolons, SentientSeven, Servel333, Sfoehner, Shamilton, Shayno, Sherool, Shirimasen, Shjacks45, Shniken1, Shop Bucuresti, SidP, Sietse Snel, Sigma 7, Silica-gel, Siliconov, Sillydragon, SimonEast, SimonP, Simulcra, Sintau.tayua, SivaKumar, Skylinerspeeder, Sleepy Sentry, Sleske, Sligocki, Smbp, Smjg, SmolderinCorpse, Smyth, Snafflekid, Snkcube, Snori, SnowRaptor, Soliloquial, Some jerk on the Internet, Someguy1221, Sonicsuns, Sonjaaa, Soren121, South Bay, SpadesSlick, Spamboy, Spartan117458, Spe88, Spik3balloon, Spike, Srijan89, Srleffler, Stacrd, Staffwaterboy, Stan Shebs, StarkRG, Ste.Ri, Ste.virus, SteinbDJ, Stephan Leeds, Stephen Gilbert, Stephen Morley, Stephenchou0722, Steven Zhang, Steveprutz, Stimson, StoneGiant, Storm Rider, Strait, Strom, Stuart P. Bentley, StuffOfInterest, Sublastic, Sully76, Sumonbd, SuperBeav, Superway25, Suvituuli, Suwa, Sven Godin, Svick, Sweeet ann, T Arndt 40, T4bits, THEN WHO WAS PHONE?, Tacvek, Tagishsimon, Tarquin, Taw, Taxman, Tbolioli, Technopat, Techtonic, Tecknode, TedE, Tedder, Thaiio, The Pondermatic, The Rambling Man, The Thing That Should Not Be, The Wild Falcon, The guy who kleanz, The.Computer1, TheDoober, TheFearow, TheJosh, Theboss48506, Thebrains29, Thechuck, Thegreatestmoever3, Theodore Kloba, Thomas Blomberg, Thorpe, ThreeBlindMice, Threyon, Thumperward, Thunderboltz, Tiefighter, TiffaF, Tim Forcer, Timaru, Timbrowne, Timeshift9, Timhoppen, Timm123, Timwi, Tiscando, Titoxd, Tjpeople, Tkgd2007, Tkteun, TobyDZ, Tobz1000, Todd Vierling, Tollsjo, Tom.freeman, Tomashcu60, Tomaxer, Tomg1234, Tomlee1968, Tommy2010, Tony1, Tooki, Torturetyler, Toyotatundra, Tpbradbury, Transfinite, Traut, Traxs7, Treygdor, TrickyNik, Truerock2, Tsedit, Tsk, Turkeyphant, TutterMouse, Tweek17, Tweisbach, Twintop, Twocs, Tymothy, UdovdM, Ugnich Anton, Uisqebaugh, UncleBubba, Undeference, UnicornTapestry, Unnyn, Unyoyega, Urhixidur, Usb3o, Useight, User5910, Utcursch, Utility Monster, Uwe Hermann, V Brian Zurita, Varnish, VasilievVV, Vaughan Pratt, Versus22, Vespristiano, Vhann, Victorgrigas, Vid, Viktor, Viljo Viitanen, Vippylaman, Vlooo, Voetsjoeba, Voidxor, Vrenator, Vuongfat, Wabernat, WackyBoots, Waiwai933, Walter Görlitz, Waltervulej, Wavelength, Wehe, Wesrick, Whitis, Whompage, WhosAsking, Whowhodilly, WikHead, Wiki alf, Wiki777777ikiw, Wikid77, Wikievil666, Wikifan21century, Wikiwinganer, Wikkrockiana, Wiknerd, Wimt, Windsok, Wizpig64, Wizzy, Wmahan, Wolbo, Wolfgang Kufner, Wonko, Woohookitty, WorldGentoo, Worm That Turned, Wtf305, Wtmitchell, Ww, WynnSmith, Wyveryx, X!, X1cygnus, Xavier86, Xezbeth, Xmm0, Xolom, Xorx, Y2kboy23, Yanayz, Yeastygoodness, Yellowdesk, Yetasoli, Yitzhak, Yosh3000, Yuckhil, Z hosen, Zac67, ZakuSage, Zalgo, Zarano, Zebe, Zenlax, Zenotek, Zephyric, Zephyris, Zidane2k1, Zippanova, Zirconscot, Zlhappyone, Zntrip, Дарко Максимовић, පසිඳු කාවින්ද, , 2619 anonymous edits

Secure Digital *Source*: http://en.wikipedia.org/w/index.php?title=Secure_Digital *Contributors*: 24frames, 81120906713, 99bluefoxx, 9allenride9, A Man In Black, AVRS, Aab298347927384, Adam Schloss, Adam2288, Adsp, Aeons, Agateller, Aikenware, Airplaneman, Akaustav, Alansohn, Aldie, AlexJ, Alexbuster500, Alexthekiwi, Allaun, Alphaman, Alvis, Amoiwangjy, Amram99, Andrevan, Andrew sh, AndrewKepert, Andrewpmk, Andru nl, Angela, Angerdan, Animehawaii, Appraiser, Aqn, Aribronstein, Armando, Arrowcatcher, Asim18, AssetBurned, Atanasov, Atompowered, Austinmurphy, Autoandragogist, Axcelis555, AzaToth, BENNYSOFT, BPM, Bastique, Bdelisle, Beao, Bearcat, Beeline23, Beland, Bergsten, Bholstege, Billgordon1099, Binba, Bjdehut, Blanchardb, BlindWanderer, Blorg, Bobblewik, Bobianite, Boboangel, Bomazi, Bookandcoffee, Bostwickenator, Brer vole, Brian Patrie, Brian-L, Briandgregory, Brianski, Brouhaha, Buster2058, Buuneko, C 1, Calcprogrammer1, Caleson, Calor, Can't sleep, clown will eat me, Canrocks, Carreon77, CaseyPenk, Caspertheghost, Catskul, Cdrum, CesarB, Chandu15, CharlotteWebb, Chealer, Chris the speller, ChrisCork, Chrisjj3, Christian75, Ckmac97, Climber22, Clorox, Cmdrjameson, Compellingelegance, Compuguy1088, Coneslayer, Consumed Crustacean, Cootiequits, Copysan, Corge, Coughinink, CrookedAsterisk, Csendesmark, Cyawman, Cybercobra, Cybertai, CyclePat, D0nj03, DMahalko, Dabomb87, Dah31, Dale Arnett, Damian Yerrick, Dan100, DanMS, Dancingonmice, Dancter, Dandv, Daniel Newby, Danielt998, Dankru, Darin-0, Darkxsun, Darrien, Daveswagon, David from Downunder, David spector, David.Monniaux, Davidswelt, Dawd, Deineka, Dennbruce, DervishD, Dhanashekar, Djm1279, Dlrohrer2003, Doc glasgow, DocWatson42, Doctorno, Dogcow, Drabant, Draconiator, Dreadstar, Drmies, Dustin Howett, Długosz, EJSawyer, Echtner, Ed Poor, Eftpotrm, Electricnet, Electron9, ElementFire, Elemesh, Elvey, Elwilke, EncMstr, Ente75, Erencexor, Ericzhang789, Esowteric, Eurleif, Evice, Excirial, Exearly, ExplicitImplicity, ExportRadical, FCartegnie, Fast healthy fish, Feedmecereal, Feydey, Flarn2006, FlashSheridan, Flasher, Fleminra, Flowerflower, Fmillour, Freddyzdead, Fresheneesz, Fudoreaper, Fullerene, Funandtrvl, Furrykef, Fvw, Gadol87, Gaselvin, Ge0rge, Geekosaurus, Gglockner, Ghaib, Glaw10, GodGell, Goosnarrggh, GraemeLeggett, Grafen, Graham87, GregorB, Grunt, HDCase, Haipa Doragon, Hankwang, Hansa11, Harlekeyn, Harvester, Hebrides, Henry W. Schmitt, Heron, Heybales, Hgrosser, Hoagg, Hooperbloob, HoserHead, Hurk87, IAmAI, IW.HG, IceHunter, Ichimonji10, IkonicDeath, Iliev, Impi, Intelliot, InternetMeme, Irayo, Ismouton, J.delanoy, JJLatWiki, Jachim, JackTinWNY, Jafeluv, Jaho, James JK, JamesAM, JamesTeterenko, Jdowland, Jeffq, Jerryobject, Jhsounds, Jidanni, Jim McKeeth, Jim.henderson, Jj2235, Jlslspam, Jnavas, Joaoplim, Jocunddus, Joe Sewell, JohnAlexan, Johnteslade, JordoCo, Jory, Josi.ow, JudyJohn, Justin Ormont, Jvr725, Jw21, Kartano, Kaszeta, Kazrak, Kbdank71, Kiimmyluff, Kinema, Kingpin13, Kjohna, Klkmiooo, KnowBuddy, Koolman2, Korj.by, Kozuch, KryptoCleric, KsprayDad, Kungfujoe, Kungming2, Kusunose, Kvng, LM1987, Lamrock, Lavenderbunny, Lbecque, Le Déchaîné, LeoO3, Leszek Jańczuk, Liaocyed, Lightmouse, LilHelpa, Lindosland, Little Professor, Littleendian, Lord Nightmare, Lordsatri, LostLeviathan, LovesMacs, Lucasreddinger, MER-C, MMuzammils, Macrakis, MadisSepp, MarXidad, Marianocecowski, Mark Renier, Martarius, Maury Markowitz, Maxim Leyenson, Maxí, Mboverload, Megazero, Memorysuppliers, Metallica10, Michael L. Kaufman,

MichaelJanich, Microsofkid, Mihai Capotă, Mikus, Misterkillboy, Mju7nhy6, Mmj, Mnts, Modster, Moreati, Mortense, Moxfyre, MrBurns, Mram80, Mrdungx, Mrzaius, Msaunier, Mwarren us, Myke2020, N1truX, N5iln, NEMT, Nagualdesign, Nahuel.carvajal, Nasukaren, Neelix, Neilka, NerdyNSK, Neurolysis, Newzack, Nil Einne, Nintendude, Nixdorf, Nnan, Noir, Noq, Notmicro, Now3d, Ntsimp, Nulltransfer, Nurg, Nv8200p, OMA2k, Oahiyeel, Oe1kenobi, Offbeatcinema, Ojigiri, Oliverdl, Omeganian, Omegatron, Oxymoron83, Pabouk, Paranoid, PatrikR, Peace81, Peaceninja28, Peter Campbell, Peter S., Peterl, Peterrobbemond, Peyre, Photographerguy, Photoguy439, Phr, Pickmynose1999, PierreOssman, Pinkevin, Piper8, Placi1982, Plasmaroo, Pranjal87, ProhibitOnions, Psharrock, Psiddall, Quaeler, Queefycreatures, Ravedave, Ravenperch, Razor2988, Rcawsey, RealGrouchy, Reaperducer, Rebroad, RichardTector, Riddley, RingtailedFox, Rjc34, Rjquillin, Rjwilmsi, Rlcantwell, Rmsuperstar99, Rmunn, Rofl cawpters, Rohan Jayasekera, RoseTech, Rosenbluh, Rstoplabe14, Ruud Koot, SD Card Version 2.Wiki.0, SF88, SGBailey, SMD915, SPKirsch, SRG275, Saimhe, Saipraneethn, Saltmiser, Sam8, SanGatiche, Sandymac, Saravanants, Sasan.j, Saverworld2, Sbmeirow, SchnitzelMannGreek, Scotty jasper2000, SeanAhern, Seaphoto, Sebadee, Sedimin, Sergei, Shadowjams, Shanes, SidP, Sin-man, Skathol, Skierpage, Skyykj, Slakr, Slashme, Slicing, Smack, Smileyborg, Smyth, Snolygoster, Speedevil, Spike-from-NH, Spikesagal, Stanleyivan, Starnestommy, Steinsomers, Stephan Leeds, Stevage, Stuston, Sunny house, Surf243, Suriel1981, Suruena, Szzuk, TAC-3, TCav, Tacvek, Tamiledu, Tarquin, The Giant Puffin, The Thing That Should Not Be, TheDoober, Themoment, ThevillagesmithE, Thewikipedian, Thunderbird2, Thunderpenguin, TimSE, Timharwoodx, Tintenfischlein, Tiredofscams, Tkgd2007, Tmansour, TobleRone, Todd Vierling, Toehead2001, Tomgibbons, Tone, Tony1, Tonyhawz, Totakeke423, Towel401, Tpbradbury, Triscal1990, Tristan Schmelcher, Twinmostech, UU, UberMan5000, UltraMagnus, Vdcappel, Versus22, Vexorg, Viggio, Viol8or, Vladsinger, VoxLuna, Waffle, Wammes Waggel, Wanderer099, Wbt92, Wbuch, Weezee, Wernher, WhiteDragon, Whitis, Wibbble, WikHead, WinTakeAll, Wine Guy, Wirbelwind, Wjejskenewr, Wjw0111, Wolfling, WriterHound, Wrlee, Wtshymanski, Wuhwuzdat, XP1, Xdddex, Xenon54, Xmhd, Xonicx, Xpclient, Yaniv Kunda, Yealout, Z hosen, Zarcillo, Zarenor, Zbrahead91, Zerpent, Ziga, Zodon, Zoicon5, Zsexdrcft, Zurotai, Zviangi, 1131 anonymous edits

PictBridge *Source*: http://en.wikipedia.org/w/index.php?title=PictBridge *Contributors*: AdSR, Ansett, Apoltix, Borisbaran, Ccacsmss, CecilWard, Cristan, Damian Yerrick, Estel, Fultron89, Ganymead, Gew, Giac83, Grstain, Hgrosser, Hub, JR98664, Jpo, Khalid hassani, Lampak, Markhurd, Morris lin, Pjrm, Plop, Quinling, R'n'B, Rixs, Ronz, RucasHost, Sietse Snel, Somebody in the WWW, Ta bu shi da yu, Thumperward, Tom-, WhosAsking, 44 anonymous edits

Audio Video Interleave *Source*: http://en.wikipedia.org/w/index.php?title=Audio_Video_Interleave *Contributors*: 10014derek, 211.216.216.xxx, AVRS, Abmac, Airconswitch, AlistairMcMillan, Alterego, Anss123, Arite, AxelBoldt, Barticus88, Beno1000, Bige1977, Bob Jonkman, Bongdentoiac, Bruce89, Bryan Derksen, C xong, CBM, Calton, Camw, Capricorn42, Carl.bunderson, Chaos583, Cielwangqin, Cielwq, Ckatz, ClariT, Courcelles, Ctjf83, CyberSkull, Cybercobra, D0odx, DARTH SIDIOUS 2, Daeron, Danieltiger45, Dizzle, Djsmithy5, DmitriyV, Dmmaus, Dogjkekf34, Don4of4, DonDiego, Drmcclainphd, Drooling Sheep, Drshoots, ELLAPRISCI, Edwardsunwiki, EliasAlucard, ElliottHoffman, EmeryD, Estevoaei, Explicit, Famfrit, Fishup2008, Fjeinca, FleetCommand, Flubeca, Frankeriksson, Frap, Frau Holle, FrenchIsAwesome, GSZaum, GTAddict, Gerbrant, Graham87, Gumbertron, Hadal, Hardwood2020, Hede2000, Howardbannister, Iphonefans2009, Iridescent, J Di, J. M., Jdthood, Jeffrey Mall, Jeffz1, Jeni, Jennavecia, JimVC3, JordoCo, Joyce23424tg, Kamasutra, Katimawan2005, Kennedy, Kku, Kuru, Law, Lexicon, Liftarn, Liquidtenmillion, LlwynogCymru, Logictheo, Loonymonkey, Lordalpha1, Lucine098ui, Lynn-LU, MER-C, Mandarax, Marcok, Materialscientist, MattCuts, Maxidrom11, Miami33139, Mike Rosoft, Mikix, Mikus, Modster, Monkeyman, Mousez, Mskadu, Nabbia, Nick1nildram, Nixdorf, Nixeagle, Norm mit, Obankston, OverlordQ, Ozzmosis, Palpalpalpal, Pastinakel, PaulBreadly, Pdelong, Polly, Prodigy96, Profoss, Pyrospirit, Quentar, R7604, RJFJR, Radical Mallard, RandomAct, RandomWalker, Rcouto, RexNL, Rich Farmbrough, Rolypolyman, Rrw, Rtdrury, Sarregouset, Scaife, Secretlondon, Sharoneditha, ShaunMacPherson, Shibohui, SophisticatedPrimate, Speedoflight, Ssfreitas, Stegop, Stevertigo, Stib, Szajd, Tajymoid, Theinfo, TimTomTom, Tomalak geretkal, ToobMug, Tuanese, Turbojet, Vanderkoip, Verne Equinox, Versageek, Waterfly99, Wellslogan, Whkoh, Wikievil666, Wikinick, Wimt, Wo.luren, Xpclient, YUL89YYZ, Yamamoto Ichiro, Yuriybrisk, Zsinj, Zundark, , , 242 anonymous edits

Image Sources, Licenses and Contributors

Image:Tescologo.svg *Source*: http://en.wikipedia.org/w/index.php?title=File:Tescologo.svg *License*: unknown *Contributors*: Estoy Aquí, Good Olfactory, J Milburn, Ohconfucius, Skier Dude, TFoxton, 2 anonymous edits

File:Increase2.svg *Source*: http://en.wikipedia.org/w/index.php?title=File:Increase2.svg *License*: unknown *Contributors*: Sarang

File:TescoHouseCheshunt.jpg *Source*: http://en.wikipedia.org/w/index.php?title=File:TescoHouseCheshunt.jpg *License*: unknown *Contributors*: Gordon Joly from London, UK

File:Tesco Kingston Park.jpg *Source*: http://en.wikipedia.org/w/index.php?title=File:Tesco_Kingston_Park.jpg *License*: unknown *Contributors*: User:Mankind 2k

File:Tescokulim1.jpg *Source*: http://en.wikipedia.org/w/index.php?title=File:Tescokulim1.jpg *License*: unknown *Contributors*: Chinesestar

File:HooverBuildingTescoEntrance.jpg *Source*: http://en.wikipedia.org/w/index.php?title=File:HooverBuildingTescoEntrance.jpg *License*: unknown *Contributors*: DRD

File:Tesco - Potters Bar - geograph.org.uk - 239291.jpg *Source*: http://en.wikipedia.org/w/index.php?title=File:Tesco_-_Potters_Bar_-_geograph.org.uk_-_239291.jpg *License*: unknown *Contributors*: Sevela.p

File:Tesco's store Evesham - geograph.org.uk - 5736.jpg *Source*: http://en.wikipedia.org/w/index.php?title=File:Tesco's_store_Evesham_-_geograph.org.uk_-_5736.jpg *License*: unknown *Contributors*: Pibwl, Pit-yacker, Sevela.p

File:Dotty Turnbull .png *Source*: http://en.wikipedia.org/w/index.php?title=File:Dotty_Turnbull_.png *License*: unknown *Contributors*: Rodhullandemu, Sick 3, Vanished 1850, 1 anonymous edits

File:Tesco, Batley.jpg *Source*: http://en.wikipedia.org/w/index.php?title=File:Tesco,_Batley.jpg *License*: unknown *Contributors*: User:Mtaylor848

Image:20040801-008-tesco.jpg *Source*: http://en.wikipedia.org/w/index.php?title=File:20040801-008-tesco.jpg *License*: unknown *Contributors*: User:Ghouston

File:TescoExpress.jpg *Source*: http://en.wikipedia.org/w/index.php?title=File:TescoExpress.jpg *License*: unknown *Contributors*: User:Rodhullandemu

File:TescoOneStop.jpg *Source*: http://en.wikipedia.org/w/index.php?title=File:TescoOneStop.jpg *License*: unknown *Contributors*: User:Rodhullandemu

File:Steady2.svg *Source*: http://en.wikipedia.org/w/index.php?title=File:Steady2.svg *License*: unknown *Contributors*: User:Tomchen1989

File:Tesco Petrol Station, St Mellons - geograph.org.uk - 92048.jpg *Source*: http://en.wikipedia.org/w/index.php?title=File:Tesco_Petrol_Station,_St_Mellons_-_geograph.org.uk_-_92048.jpg *License*: unknown *Contributors*: Mtaylor848

Image:TescoTechLogo 03.png *Source*: http://en.wikipedia.org/w/index.php?title=File:TescoTechLogo_03.png *License*: unknown *Contributors*: AlexJFox, Good Olfactory

File:Flag of the People's Republic of China.svg *Source*: http://en.wikipedia.org/w/index.php?title=File:Flag_of_the_People's_Republic_of_China.svg *License*: unknown *Contributors*: User:Denelson83, User:SKopp, User:Shizhao, User:Zscout370

File:Flag of the Czech Republic.svg *Source*: http://en.wikipedia.org/w/index.php?title=File:Flag_of_the_Czech_Republic.svg *License*: unknown *Contributors*: special commission (of code): SVG version by cs:-xfi-. Colors according to Appendix No. 3 of czech legal Act 3/1993. cs:Zirland.

File:Flag of Hungary.svg *Source*: http://en.wikipedia.org/w/index.php?title=File:Flag_of_Hungary.svg *License*: unknown *Contributors*: User:SKopp

File:Flag of Ireland.svg *Source*: http://en.wikipedia.org/w/index.php?title=File:Flag_of_Ireland.svg *License*: unknown *Contributors*: User:SKopp

File:Flag of Japan.svg *Source*: http://en.wikipedia.org/w/index.php?title=File:Flag_of_Japan.svg *License*: unknown *Contributors*: Anomie

File:Decrease2.svg *Source*: http://en.wikipedia.org/w/index.php?title=File:Decrease2.svg *License*: unknown *Contributors*: User:Sarang

File:Flag of Malaysia.svg *Source*: http://en.wikipedia.org/w/index.php?title=File:Flag_of_Malaysia.svg *License*: unknown *Contributors*: User:SKopp

File:Flag of Poland.svg *Source*: http://en.wikipedia.org/w/index.php?title=File:Flag_of_Poland.svg *License*: unknown *Contributors*: Anomie, Mifter

File:Flag of Slovakia.svg *Source*: http://en.wikipedia.org/w/index.php?title=File:Flag_of_Slovakia.svg *License*: unknown *Contributors*: User:SKopp

File:Flag of South Korea.svg *Source*: http://en.wikipedia.org/w/index.php?title=File:Flag_of_South_Korea.svg *License*: unknown *Contributors*: Various

File:Flag of Thailand.svg *Source*: http://en.wikipedia.org/w/index.php?title=File:Flag_of_Thailand.svg *License*: unknown *Contributors*: User:Zscout370

File:Flag of Turkey.svg *Source*: http://en.wikipedia.org/w/index.php?title=File:Flag_of_Turkey.svg *License*: unknown *Contributors*: User:Dbenbenn

File:Flag of the United Kingdom.svg *Source*: http://en.wikipedia.org/w/index.php?title=File:Flag_of_the_United_Kingdom.svg *License*: unknown *Contributors*: Anomie, Mifter

File:Flag of the United States.svg *Source*: http://en.wikipedia.org/w/index.php?title=File:Flag_of_the_United_States.svg *License*: unknown *Contributors*: Anomie

File:Tesco express prague - belehradska street.jpg *Source*: http://en.wikipedia.org/w/index.php?title=File:Tesco_express_prague_-_belehradska_street.jpg *License*: unknown *Contributors*: User:Kucmel007

File:Tesco Kőszeg.JPG *Source*: http://en.wikipedia.org/w/index.php?title=File:Tesco_Kőszeg.JPG *License*: unknown *Contributors*: Pan Peter12

Image:Bukit-tinggi-klang-tesco.gif *Source*: http://en.wikipedia.org/w/index.php?title=File:Bukit-tinggi-klang-tesco.gif *License*: unknown *Contributors*: Original uploader was Klangboy at en.wikipedia

Image:POL Prokocim Tesco.jpg *Source*: http://en.wikipedia.org/w/index.php?title=File:POL_Prokocim_Tesco.jpg *License*: unknown *Contributors*: Users Vindicator, Paulka on pl.wikipedia

File:Tesco Kamenné námestie.jpg *Source*: http://en.wikipedia.org/w/index.php?title=File:Tesco_Kamenné_námestie.jpg *License*: unknown *Contributors*: Wizzard

File:Tesco-Lotus.jpg *Source*: http://en.wikipedia.org/w/index.php?title=File:Tesco-Lotus.jpg *License*: unknown *Contributors*: User:Matttes

Image:Fresh & Easy Las Vegas, Sun City Summerlin.jpg *Source*: http://en.wikipedia.org/w/index.php?title=File:Fresh_&_Easy_Las_Vegas,_Sun_City_Summerlin.jpg *License*: unknown *Contributors*: Uforik at en.wikipedia

Image:Tesco Market Share.jpg *Source*: http://en.wikipedia.org/w/index.php?title=File:Tesco_Market_Share.jpg *License*: unknown *Contributors*: User:Drunkenmonkey. Original uploader was Drunkenmonkey at en.wikipedia

File:Tesco, Burscough.JPG *Source*: http://en.wikipedia.org/w/index.php?title=File:Tesco,_Burscough.JPG *License*: unknown *Contributors*: User:Small-town hero

File:Baldock Tesco.jpg *Source*: http://en.wikipedia.org/w/index.php?title=File:Baldock_Tesco.jpg *License*: unknown *Contributors*: User:Traveler100

File:Tesco's Superstore, Swindon - geograph.org.uk - 118815.jpg *Source*: http://en.wikipedia.org/w/index.php?title=File:Tesco's_Superstore,_Swindon_-_geograph.org.uk_-_118815.jpg *License*: unknown *Contributors*: Man vyi, Sevela.p

File:Praha, Národní třída, Tesco.jpg *Source*: http://en.wikipedia.org/w/index.php?title=File:Praha,_Národní_třída,_Tesco.jpg *License*: unknown *Contributors*: User:Aktron

File:Tesco headquarters Czech.jpg *Source*: http://en.wikipedia.org/w/index.php?title=File:Tesco_headquarters_Czech.jpg *License*: unknown *Contributors*: User:Kucmel007

File:Tesco in Brno.jpg *Source*: http://en.wikipedia.org/w/index.php?title=File:Tesco_in_Brno.jpg *License*: unknown *Contributors*: User:Sevela.p

File:Tesco Mikulov.jpg *Source*: http://en.wikipedia.org/w/index.php?title=File:Tesco_Mikulov.jpg *License*: unknown *Contributors*: Original uploader was KRAX at cs.wikipedia

File:Tesco Express - geograph.org.uk - 230620.jpg *Source*: http://en.wikipedia.org/w/index.php?title=File:Tesco_Express_-_geograph.org.uk_-_230620.jpg *License*: unknown *Contributors*: Sevela.p

File:Northallerton Tesco.JPG *Source*: http://en.wikipedia.org/w/index.php?title=File:Northallerton_Tesco.JPG *License*: unknown *Contributors*: User:Chrisloader

File:Tesco Zličín.jpg *Source*: http://en.wikipedia.org/w/index.php?title=File:Tesco_Zličín.jpg *License*: unknown *Contributors*: User:ŠJů

File:Tesco Metro Oxford.jpg *Source*: http://en.wikipedia.org/w/index.php?title=File:Tesco_Metro_Oxford.jpg *License*: unknown *Contributors*: User:Chrisloader

File:Pixel-example.png *Source*: http://en.wikipedia.org/w/index.php?title=File:Pixel-example.png *License*: unknown *Contributors*: User:ed_g2s

File:Closeup of pixels.JPG *Source*: http://en.wikipedia.org/w/index.php?title=File:Closeup_of_pixels.JPG *License*: unknown *Contributors*: User:Kprateek88

Image:ReconstructionsFromPixels.png *Source*: http://en.wikipedia.org/w/index.php?title=File:ReconstructionsFromPixels.png *License*: unknown *Contributors*: Original uploader was Dicklyon at en.wikipedia

Image:Wikipedia ClearType.png *Source*: http://en.wikipedia.org/w/index.php?title=File:Wikipedia_ClearType.png *License*: unknown *Contributors*: Cy21, Hautala, Kenmayer, MGA73

Image:Pixel geometry 01 Pengo.jpg *Source*: http://en.wikipedia.org/w/index.php?title=File:Pixel_geometry_01_Pengo.jpg *License*: unknown *Contributors*: User:Pengo

Image:Canon PowerShot A95 - front and back.jpg *Source*: http://en.wikipedia.org/w/index.php?title=File:Canon_PowerShot_A95_-_front_and_back.jpg *License*: unknown *Contributors*: User:Fir0002, User:Matt57

File:Sony DSC-W170.jpg *Source*: http://en.wikipedia.org/w/index.php?title=File:Sony_DSC-W170.jpg *License*: unknown *Contributors*: Pierre Bauduin

File:Sony DSC-H2 01.jpg *Source*: http://en.wikipedia.org/w/index.php?title=File:Sony_DSC-H2_01.jpg *License*: unknown *Contributors*: User:Tschaensky

File:E-30-Cutmodel.jpg *Source*: http://en.wikipedia.org/w/index.php?title=File:E-30-Cutmodel.jpg *License*: unknown *Contributors*: User:Hanabi123

Image:Matakis - blurred.jpg *Source*: http://en.wikipedia.org/w/index.php?title=File:Matakis_-_blurred.jpg *License*: unknown *Contributors*: User:Mikael Häggström

Image:MARTAKIS1.jpg *Source*: http://en.wikipedia.org/w/index.php?title=File:MARTAKIS1.jpg *License*: unknown *Contributors*: User:Bratopoulosm

Image:Hendys Law.jpg *Source*: http://en.wikipedia.org/w/index.php?title=File:Hendys_Law.jpg *License*: unknown *Contributors*: Original uploader was Barry.hendy at en.wikipedia

Image:Ccd-sensor.jpg *Source*: http://en.wikipedia.org/w/index.php?title=File:Ccd-sensor.jpg *License*: unknown *Contributors*: User:C-M

Image:Partly disassembled Lumix digital camera.jpg *Source*: http://en.wikipedia.org/w/index.php?title=File:Partly_disassembled_Lumix_digital_camera.jpg *License*: unknown *Contributors*: Steve Jurvetson from Menlo Park, USA

Image:Bayer pattern on sensor.svg *Source*: http://en.wikipedia.org/w/index.php?title=File:Bayer_pattern_on_sensor.svg *License*: unknown *Contributors*: 32bitmaschine, AnonMoos, Cburnett, W!B:

Image:kids 50mm 100mm.jpg *Source*: http://en.wikipedia.org/w/index.php?title=File:Kids_50mm_100mm.jpg *License*: unknown *Contributors*: Original uploader was Toytoy at en.wikipedia

Image:SensorSizes.svg *Source*: http://en.wikipedia.org/w/index.php?title=File:SensorSizes.svg *License*: unknown *Contributors*: User:Hotshot977

Image:Cf-card.jpg *Source*: http://en.wikipedia.org/w/index.php?title=File:Cf-card.jpg *License*: unknown *Contributors*: Original uploader was Motorrad-67 at en.wikipedia

Image:Mini CD vs Normal CD comparison.jpg *Source*: http://en.wikipedia.org/w/index.php?title=File:Mini_CD_vs_Normal_CD_comparison.jpg *License*: unknown *Contributors*: Original uploader was EddieBernard at en.wikipedia

Image:MicroDrive1GB.jpg *Source*: http://en.wikipedia.org/w/index.php?title=File:MicroDrive1GB.jpg *License*: unknown *Contributors*: User:Afrank99

Image:USB flash drive.jpg *Source*: http://en.wikipedia.org/w/index.php?title=File:USB_flash_drive.jpg *License*: unknown *Contributors*: User:Dori

Image:Floppy disk 90mm.JPG *Source*: http://en.wikipedia.org/w/index.php?title=File:Floppy_disk_90mm.JPG *License*: unknown *Contributors*: Afrank99, BrokenSegue, Jeandré du Toit

File:Woman selling photo stuff.jpg *Source*: http://en.wikipedia.org/w/index.php?title=File:Woman_selling_photo_stuff.jpg *License*: unknown *Contributors*: Eric Ward from Provo, UT, USA

File:Reflex camera numeric.svg *Source*: http://en.wikipedia.org/w/index.php?title=File:Reflex_camera_numeric.svg *License*: unknown *Contributors*: User:Jean François WITZ

File:Energizer reghargeble batteryIMG 0006.JPG *Source*: http://en.wikipedia.org/w/index.php?title=File:Energizer_reghargeble_batteryIMG_0006.JPG *License*: unknown *Contributors*: Kookyunii, MGA73, 1 anonymous edits

File:6 most common battery types-1.jpg *Source*: http://en.wikipedia.org/w/index.php?title=File:6_most_common_battery_types-1.jpg *License*: unknown *Contributors*: User:Lead holder

File:Solar AA charger 01 Pengo.jpg *Source*: http://en.wikipedia.org/w/index.php?title=File:Solar_AA_charger_01_Pengo.jpg *License*: unknown *Contributors*: User:Pengo

Image:Nikkor 28-200 zoom.jpg *Source*: http://en.wikipedia.org/w/index.php?title=File:Nikkor_28-200_zoom.jpg *License*: unknown *Contributors*: Original uploader was Marc Lacoste at en.wikipedia

Image:Zoom effect.jpg *Source*: http://en.wikipedia.org/w/index.php?title=File:Zoom_effect.jpg *License*: unknown *Contributors*: Daniil Moerman (= user B166ER)

Image:Zoomar MG 1895a.jpg *Source*: http://en.wikipedia.org/w/index.php?title=File:Zoomar_MG_1895a.jpg *License*: unknown *Contributors*: User:Rama

Image:Zoomlens1.svg *Source*: http://en.wikipedia.org/w/index.php?title=File:Zoomlens1.svg *License*: unknown *Contributors*: User:Jxjl

Image:Zoomlens2.svg *Source*: http://en.wikipedia.org/w/index.php?title=File:Zoomlens2.svg *License*: unknown *Contributors*: User:Jxjl

Image:Zoom prinzip.gif *Source*: http://en.wikipedia.org/w/index.php?title=File:Zoom_prinzip.gif *License*: unknown *Contributors*: User:Smial

Image:Focal-length.svg *Source*: http://en.wikipedia.org/w/index.php?title=File:Focal-length.svg *License*: unknown *Contributors*: User:Henrik

File:ThickLens.png *Source*: http://en.wikipedia.org/w/index.php?title=File:ThickLens.png *License*: unknown *Contributors*: User:Tamasflex

Image:Angleofview 28mm f4.jpg *Source*: http://en.wikipedia.org/w/index.php?title=File:Angleofview_28mm_f4.jpg *License*: unknown *Contributors*: User:Koyaanis Qatsi

Image:Angleofview 50mm f4.jpg *Source*: http://en.wikipedia.org/w/index.php?title=File:Angleofview_50mm_f4.jpg *License*: unknown *Contributors*: User:Koyaanis Qatsi

Image:Angleofview 70mm f4.jpg *Source*: http://en.wikipedia.org/w/index.php?title=File:Angleofview_70mm_f4.jpg *License*: unknown *Contributors*: User:Koyaanis Qatsi

Image:Angleofview 210mm f4.jpg *Source*: http://en.wikipedia.org/w/index.php?title=File:Angleofview_210mm_f4.jpg *License*: unknown *Contributors*: User:Koyaanis Qatsi

Image:Thin_lens_images.svg *Source*: http://en.wikipedia.org/w/index.php?title=File:Thin_lens_images.svg *License*: unknown *Contributors*: User:Cmglee

Image:Usbcardreader.png *Source*: http://en.wikipedia.org/w/index.php?title=File:Usbcardreader.png *License*: unknown *Contributors*: Alexander Jones

file:Certified USB.svg *Source*: http://en.wikipedia.org/w/index.php?title=File:Certified_USB.svg *License*: unknown *Contributors*: KUsam, Zyxw, 1 anonymous edits

file:USB.svg *Source*: http://en.wikipedia.org/w/index.php?title=File:USB.svg *License*: unknown *Contributors*: User:LivingShadow

file:USB Icon.svg *Source*: http://en.wikipedia.org/w/index.php?title=File:USB_Icon.svg *License*: unknown *Contributors*: Mobius at en.wikipedia

File:USB-Connector-Standard.jpg *Source*: http://en.wikipedia.org/w/index.php?title=File:USB-Connector-Standard.jpg *License*: unknown *Contributors*: User:Evan-Amos

File:Computer-USB2-card.jpg *Source*: http://en.wikipedia.org/w/index.php?title=File:Computer-USB2-card.jpg *License*: unknown *Contributors*: User:Evan-Amos

file:Certified Hi-Speed USB.svg *Source*: http://en.wikipedia.org/w/index.php?title=File:Certified_Hi-Speed_USB.svg *License*: unknown *Contributors*: Frap, KUsam, Zyxw

file:SuperSpeed USB.svg *Source*: http://en.wikipedia.org/w/index.php?title=File:SuperSpeed_USB.svg *License*: unknown *Contributors*: Caspertheghost

file:USB pipes and endpoints (en).svg *Source*: http://en.wikipedia.org/w/index.php?title=File:USB_pipes_and_endpoints_(en).svg *License*: unknown *Contributors*: User:Bdesham

file:USB Front Port.jpg *Source*: http://en.wikipedia.org/w/index.php?title=File:USB_Front_Port.jpg *License*: unknown *Contributors*: User:Wyveryx

file:SanDisk Cruzer Micro.png *Source*: http://en.wikipedia.org/w/index.php?title=File:SanDisk_Cruzer_Micro.png *License*: unknown *Contributors*: User:Beao, User:Evan-Amos

file:Types-usb new.svg *Source*: http://en.wikipedia.org/w/index.php?title=File:Types-usb_new.svg *License*: unknown *Contributors*: User:Darx

file:USB 3.0 Micro B plug.PNG *Source*: http://en.wikipedia.org/w/index.php?title=File:USB_3.0_Micro_B_plug.PNG *License*: unknown *Contributors*: User:Tosaka

file:Male and Female USB Connectors.jpg *Source*: http://en.wikipedia.org/w/index.php?title=File:Male_and_Female_USB_Connectors.jpg *License*: unknown *Contributors*: Original uploader was Zephyris at en.wikipedia Later version(s) were uploaded by Osama bin dipesh at en.wikipedia.

file:Usb extension cable.jpg *Source*: http://en.wikipedia.org/w/index.php?title=File:Usb_extension_cable.jpg *License*: unknown *Contributors*: Original uploader was J.smith at en.wikipedia

file:Usb connectors.JPG *Source*: http://en.wikipedia.org/w/index.php?title=File:Usb_connectors.JPG *License*: unknown *Contributors*: User:Viljo Viitanen

File:MicroB USB Plug.jpg *Source*: http://en.wikipedia.org/w/index.php?title=File:MicroB_USB_Plug.jpg *License*: unknown *Contributors*: User:masamic

File:Mini usb AB.jpg *Source*: http://en.wikipedia.org/w/index.php?title=File:Mini_usb_AB.jpg *License*: unknown *Contributors*: User Mgdunn on en.wikipedia

File:USB Std A.png *Source*: http://en.wikipedia.org/w/index.php?title=File:USB_Std_A.png *License*: unknown *Contributors*: User:Music Sorter

File:USB Std B.png *Source*: http://en.wikipedia.org/w/index.php?title=File:USB_Std_B.png *License*: unknown *Contributors*: User:Music Sorter

File:USB Mini B.png *Source*: http://en.wikipedia.org/w/index.php?title=File:USB_Mini_B.png *License*: unknown *Contributors*: User:Music Sorter

File:USB Micro A.png *Source*: http://en.wikipedia.org/w/index.php?title=File:USB_Micro_A.png *License*: unknown *Contributors*: User:Music Sorter

File:USB Micro B.png *Source*: http://en.wikipedia.org/w/index.php?title=File:USB_Micro_B.png *License*: unknown *Contributors*: User:Music Sorter

File:USB Mini-B receptacle.png *Source*: http://en.wikipedia.org/w/index.php?title=File:USB_Mini-B_receptacle.png *License*: unknown *Contributors*: User:Music Sorter

File:USB Micro-AB receptacle.jpg *Source*: http://en.wikipedia.org/w/index.php?title=File:USB_Micro-AB_receptacle.jpg *License*: unknown *Contributors*: User:Music Sorter

File:USB Micro-B receptacle.jpg *Source*: http://en.wikipedia.org/w/index.php?title=File:USB_Micro-B_receptacle.jpg *License*: unknown *Contributors*: User:Music Sorter

file:Ext usb.gif *Source*: http://en.wikipedia.org/w/index.php?title=File:Ext_usb.gif *License*: unknown *Contributors*: Groink, Julienrl

file:Popport.jpg *Source*: http://en.wikipedia.org/w/index.php?title=File:Popport.jpg *License*: unknown *Contributors*: w:en:User:EiZeiEiZei

file:USB Twisted Pair.svg *Source*: http://en.wikipedia.org/w/index.php?title=File:USB_Twisted_Pair.svg *License*: unknown *Contributors*: User:Inductiveload, User:WolfWings

file:Micro USB phone charger.jpg *Source*: http://en.wikipedia.org/w/index.php?title=File:Micro_USB_phone_charger.jpg *License*: unknown *Contributors*: User:Reinraum

file:USBVacuumCleaner.jpg *Source*: http://en.wikipedia.org/w/index.php?title=File:USBVacuumCleaner.jpg *License*: unknown *Contributors*: User:Raysonho

File:SD Cards.svg *Source*: http://en.wikipedia.org/w/index.php?title=File:SD_Cards.svg *License*: unknown *Contributors*: Tkgd2007,

File:SD-microSD adaptor.jpg *Source*: http://en.wikipedia.org/w/index.php?title=File:SD-microSD_adaptor.jpg *License*: unknown *Contributors*: Yoskov

File:MicroSD MemoryCard 002.jpg *Source*: http://en.wikipedia.org/w/index.php?title=File:MicroSD_MemoryCard_002.jpg *License*: unknown *Contributors*: User:Kropsoq

File:TwinMOS 16gb sdhc.jpg *Source*: http://en.wikipedia.org/w/index.php?title=File:TwinMOS_16gb_sdhc.jpg *License*: unknown *Contributors*: User:Z hosen

File:HP PhotoSmart SDIO Kamera.jpg *Source*: http://en.wikipedia.org/w/index.php?title=File:HP_PhotoSmart_SDIO_Kamera.jpg *License*: unknown *Contributors*: User:Afrank99

File:USB-SD-Cards.jpg *Source*: http://en.wikipedia.org/w/index.php?title=File:USB-SD-Cards.jpg *License*: unknown *Contributors*: User:AssetBurned

File:Colors of SD card.jpg *Source*: http://en.wikipedia.org/w/index.php?title=File:Colors_of_SD_card.jpg *License*: unknown *Contributors*: User:Spike-from-NH

File:SDHC HD C16 32GB .jpg *Source*: http://en.wikipedia.org/w/index.php?title=File:SDHC_HD_C16_32GB_.jpg *License*: unknown *Contributors*: User:ACGFAN

File:SDHC Speed Class 2.svg *Source*: http://en.wikipedia.org/w/index.php?title=File:SDHC_Speed_Class_2.svg *License*: unknown *Contributors*: User:Kizar

File:SDHC Speed Class 4.svg *Source*: http://en.wikipedia.org/w/index.php?title=File:SDHC_Speed_Class_4.svg *License*: unknown *Contributors*: User:Kizar

File:SDHC Speed Class 6.svg *Source*: http://en.wikipedia.org/w/index.php?title=File:SDHC_Speed_Class_6.svg *License*: unknown *Contributors*: User:Kizar

File:SDHC Speed Class 10.svg *Source*: http://en.wikipedia.org/w/index.php?title=File:SDHC_Speed_Class_10.svg *License*: unknown *Contributors*: User:Kizar

File:Sd4gy crop.jpg *Source*: http://en.wikipedia.org/w/index.php?title=File:Sd4gy_crop.jpg *License*: unknown *Contributors*: User:Wirepath

File:Canon hf100 with memory card.jpg *Source*: http://en.wikipedia.org/w/index.php?title=File:Canon_hf100_with_memory_card.jpg *License*: unknown *Contributors*: User:Mikus

File:Arduino Ethernet Shield (pre-production sample).jpg *Source*: http://en.wikipedia.org/w/index.php?title=File:Arduino_Ethernet_Shield_(pre-production_sample).jpg *License*: unknown *Contributors*: Matt Biddulph

File:8 bytes vs. 8Gbytes.jpg *Source*: http://en.wikipedia.org/w/index.php?title=File:8_bytes_vs._8Gbytes.jpg *License*: unknown *Contributors*: Daniel Sancho from Málaga, Spain

File:Sdadaptersandcards.jpg *Source*: http://en.wikipedia.org/w/index.php?title=File:Sdadaptersandcards.jpg *License*: unknown *Contributors*: Original uploader was Atanasov at en.wikipedia Later version(s) were uploaded by Brybry26 at en.wikipedia.

File:Flash memory cards size.jpg *Source*: http://en.wikipedia.org/w/index.php?title=File:Flash_memory_cards_size.jpg *License*: unknown *Contributors*: Afrank99, Ivob, MMuzammils, Moxfyre, Rhe br, Solomon203, Warden, Zxb, 1 anonymous edits

File:Multiple memory card reader writer.jpg *Source*: http://en.wikipedia.org/w/index.php?title=File:Multiple_memory_card_reader_writer.jpg *License*: unknown *Contributors*: User:Qurren

File:SD card pinning.jpg *Source*: http://en.wikipedia.org/w/index.php?title=File:SD_card_pinning.jpg *License*: unknown *Contributors*: User:Afrank99, User:Jaho

File:Secure digital card usb-adapter 3.jpg *Source*: http://en.wikipedia.org/w/index.php?title=File:Secure_digital_card_usb-adapter_3.jpg *License*: unknown *Contributors*: User:Hundehalter

File:Sd insides.png *Source*: http://en.wikipedia.org/w/index.php?title=File:Sd_insides.png *License*: unknown *Contributors*: User:Saltmiser

File:SD-extreMEmory 2GB alt innen.jpg *Source*: http://en.wikipedia.org/w/index.php?title=File:SD-extreMEmory_2GB_alt_innen.jpg *License*: unknown *Contributors*: User:Manorainjan

File:SDHC inside.jpg *Source*: http://en.wikipedia.org/w/index.php?title=File:SDHC_inside.jpg *License*: unknown *Contributors*: User:George Shuklin

File:Pictbridge symbol.png *Source*: http://en.wikipedia.org/w/index.php?title=File:Pictbridge_symbol.png *License*: unknown *Contributors*: User:Andrewpmk, User:BetacommandBot, User:Cydebot, User:Luis Dantas, User:Plm209, User:Polbot, User:Somebody in the WWW

MIX
Papier aus verantwortungsvollen Quellen
Paper from responsible sources
FSC® C105338

FSC
www.fsc.org

Printed by Books on Demand GmbH, Norderstedt / Germany